Wetland Plants of the Adirondacks: Ferns, Woody Plants, and Graminoids

Meiyin Wu & Dennis Kalma

Order this book online at www.trafford.com
or email orders@trafford.com

Most Trafford titles are also available at major online book retailers.

Printed in the United States of America.

ISBN: 978-1-4269-5840-3 (sc)
ISBN: 978-1-4269-6057-4 (hc)
ISBN: 978-1-4269-5841-0 (e)

Library of Congress Control Number: 2011902741

Trafford rev. 02/24/2011

www.trafford.com

North America & International
toll-free: 1 888 232 4444 (USA & Canada)
phone: 250 383 6864 • fax: 812 355 4082

Table of Contents

Acknowledgements

We want to thank agencies that provided funding and/or support: the United States Environmental Protection Agency, the United States Department of Agriculture Natural Resources Conservation Service the New York State Department of Environmental Conservation, the New York State Adirondack Park Agency, the Nature Conservancy, Ausable River Association, Boquet River Association, State University of New York College at Plattsburgh, and Montclair State University. Much appreciation goes to the many landowners who have allowed us access to the wetlands on their lands. Special thanks to Dr. Kenneth Adams and Mr. Dan Spada who shared their knowledge and expertise. We thank Mr. Gabriel Jimenez for his technical assistance. A final note of deepest love and gratitude goes to our families—Alice, Allen, Eddie, Michael, and Meiyin's parents, whom without their love, patience and support, this book could never be finished.

Introduction

This book is the first of two books on identification of wetland plants in the Adirondack region. Due to page limitation, we cannot possibly include every wetland plant in the Adirondack region; we tried to include the most common ones. Although not true wetland plants, some species are commonly found within Adirondack wetlands; therefore, those species were selected to be included intentionally. The two books begin by splitting the species into five major groups:

Book 1:
- Ferns and Allies
- Woody Plants
- Grass-like Plants

Book 2:
- Herbaceous Plants
- Aquatic Plants

While some species might be easy to identify, others may belong to a group of species where identification can be a difficult technical task and requires specific botanical information. Every effort has been made to minimize the amount of terminology used in these books. Where possible, the rather bewildering descriptive terms used in technical manuals have been replaced by plain English translations. We hope these two books will be useful to anyone without botanical training. Some familiarity with the names of parts of plants is, however, necessary. The following section gives a minimal introduction to the structure of plants and the terminology involved to describe them. Some specialized structures, found only in certain groups of plants, are described in the introduction to those groups.

Identification by Field Guide versus Keys:

Some species are so distinctive that it is a waste of time to construct and use a key to distinguish among them. For example, a jack-in-the-pulpit is almost impossible to confuse with anything else in Adirondack wetlands, so it makes sense to just look at the individual accounts of the species in the subgroup the general key leads you to and choose among them. Other species are more difficult (for example, the sedges) and have such a mix characteristics that they are difficult to distinguish among without a key. In these cases a key will be helpful and will lead you directly to the identity of a species. Or in some cases, the key will lead you to a list of species; you can then look at the individual accounts of these species and determine the appropriate one from among them.

Keys:

A key is like the game "Twenty Questions". At each level you are asked to choose between two answers to a question or between two statements. Depending on your choice, you are led to other questions. Each question in turn eliminates some possible species. After a series of questions a final question leads you to the name of the species (or a group of species).

In these books there are always two, and only two, possible choices at each level of the key. The alternate choices are called a "couplet". At the left hand side of the page is a number followed by "a" or "b" that identifies the couplet, e.g. 1a) & 1b). Each half of the couplet is followed by statements between which you must choose. At the end of the line of each half of the couplet is a number that points you to the next set of couplets to choose between. If the choice in the couplet is the end point of that branch of the key, then the line ends in the name of the species or in the name of a group to which you proceed for another key. As an example, for the identification of the major groups, as described above, a very simple key would be constructed as follows:

1a) Plants simple: fern-like or horsetail or quillworts... Ferns & Allies (*Go to Book 1 page 7*)
1b) Others ... 2

2a) Plants with woody stems ... Woody Plants (*Go to Book 1 page 23*)
2b) Plants without woody stems ... 3

3a) Plants with grass-like stems and leaves ... Grass-like Plants (*Go to Book 1 page 110*)
3b) Plants not grass-like ... 4

4a) Plants with leaves entirely underwater, floating on the surface, or plants free-floating ... Aquatic Plants (*Go to Book 2 page 149*)
4b) Plants without leaves entirely underwater, floating on the surface, or plants free-floating ... Herbaceous Plants (*Go to Book 2 page 7*)

Flower Anatomy:

The structure of the flower is probably the most important characteristic biologists use to determine the evolutionary relationships among the plants. Because of this most keys for identification rely heavily on the flowers. Although they may not always be available for use, field biologists usually consider them the "gold standard" for identification.

The structure of a typical flower is shown in the illustration:

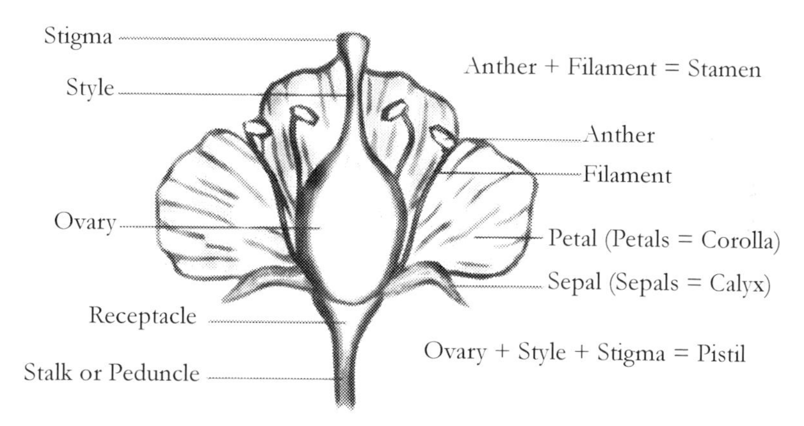

The ovary, stigma, and style, the female reproductive parts, are collectively known as the pistil. The filament and anther, the male reproductive parts, are collectively known as the stamen. The accessory structures, the petals (collectively known as the corolla) and the sepals (collectively known as the calyx) may or may not be present; the petals and sepals together are known as the perianth or, in some groups, as the tepals. The flower may have radial symmetry (the symmetry of a circle – looking down from the top, the sides are the same no matter where you divide it) or may have bilateral symmetry (there are definite right and left sides, same as with our own bodies). Different keys use different words for these terms; for example radial symmetry is often known as "rotate" or "symmetrical", while bilateral symmetry is often known as "zygomorphic".

The petals or the sepals may fuse all along their length, or along part of their length, to give a tubular corolla or tubular calyx. In the diagram above the ovary sits on top of the receptacle – a superior ovary. The receptacle may grow up and around the ovary so it appears as if the calyx and corolla are attached above the ovary – an inferior ovary. In many flowers, such as the one pictured here, the female (pistil) and male (stamen) parts are found in the same flower. These flowers are considered perfect. In other cases, imperfect or unisexual flowers, the male parts are found in some flowers, staminate flowers, and the female parts are found in other flowers, pistillate flowers. Staminate and pistillate flowers may be found on the same plant (dioecious plants) or they may be found on different plants (monoecious plants).

Flowers may be single or they may be grouped in inflorescences. The various arrangements have different names:

1. **Spike**: In a spike the the individual flowers are attached directly to a central stem (the rachis).
2. **Raceme**: A raceme is similar to a spike but the individual flowers are at the ends of side branches of the rachis.
3. **Panicle**: A panicle is similar to a raceme but the side branches are themselves branched.
4. **Corymb**: A corymb is like a raceme but the length of the branches varies so that all of the flowers are in a more or less flat-topped inflorescence. In a compound corymb each of the side branches is branched.
5. **Umbel**: In an umbel all of the flower stalks arise at a single point at the top of the stem. The inflorescence may be flat-topped or rounded. In a compound umbel each of the branches is branched.
6. **Head**: In a head all of the flowers arise from a small "head".

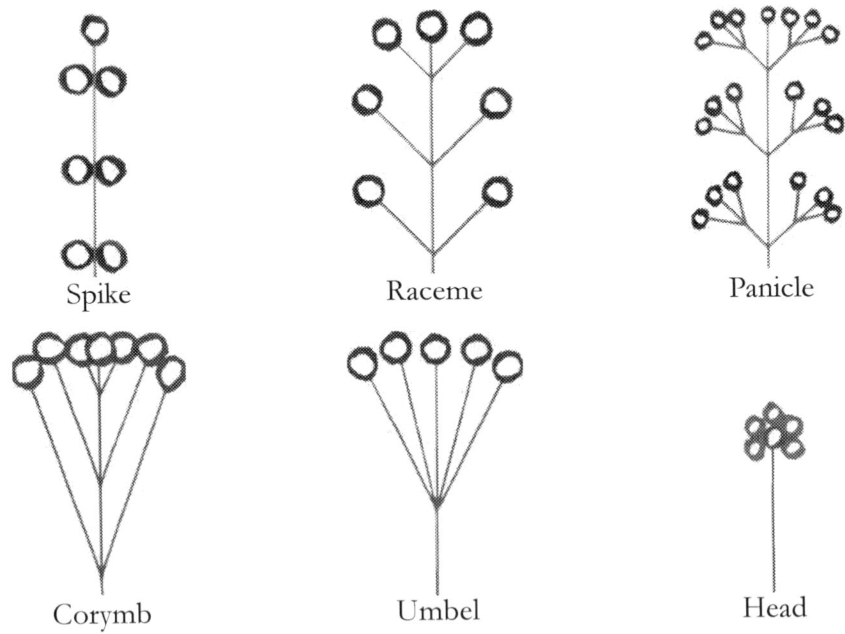

Vegetative Structures:

The leaves and their arrangement are also used in identification. Some perennial plants have underground rhizome systems which extend horizontally and send out new roots and shoots year after year. Parasitic plants may have "No Leaves". In a few other plants the leaves may disappear either before or after the flowers are present; these are referred to as "No Leaves At Flowering". When leaves are present, they may arise directly from the ground, "Leaves Basal", or they may be on the stem. In some herbaceous plants the distinction of whether the leaves are truly basal or whether they just arise near the base of the stem is problematic; in these plants we have placed the species in both groupings – whichever choice you make should lead you to the correct answer. If they are on the stem they may be arranged in pairs or whorls opposite one another, "Leaves Opposite or Whorled", or they may be arranged one after the other on alternate sides of the stem, "Leaves Alternate". The leaves may consist of a single undivided blade, "Leaves Simple", or the blade may be subdivided into leaflets, "Leaves Compound". The leaf (or leaflet) may have a smooth edge ("Entire") or it may have teeth ("Toothed") or it may have lobes ("Lobed"). If an indentation in the blade of the leaf extends only part way to the central vein, the leaf considered is lobed; if the indentation extends to the central vein, so that there is a stretch of vein without blade tissue along it, the leaf is subdivided into leaflets and is considered compound.

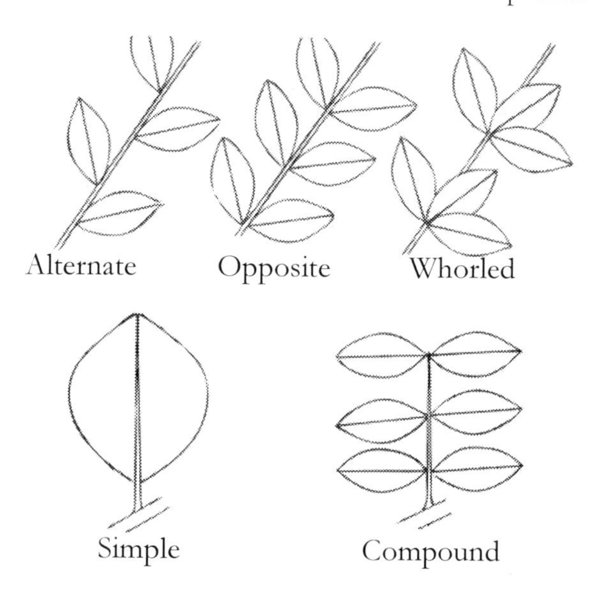

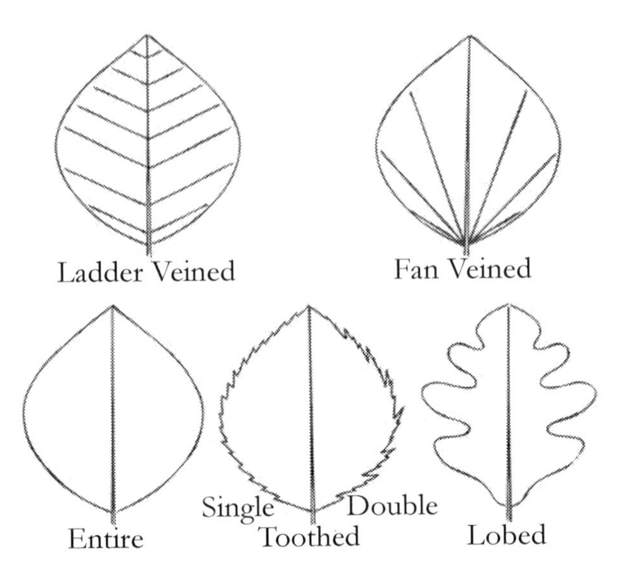

Collecting Specimens:

If a species cannot be identified in the field, specimens might be collected to verify identifications. Particularly in the case of grasses-like species and certain other groups, where identification of species is often difficult and depends on microscopic characters. When collecting specimens of a tree, collect a short segment of branch with attached leaves from and document its height, the texture of its bark, and so on. When collecting specimens of the grass-like species, care should be taken to collect a portion of the roots as well as the stems and seed heads as these are often critical characters. Specimens collected should be placed in plastic bags and kept in a cool location (refrigerate if necessary to hold overnight). If the specimen seems to be rare it should not be collected. Many biologists use the "rule of 20" as a guideline for collecting unidentified specimens: if there are more than 20 plants in the area it is appropriate to collect one for identification. Orchids, however, should never be collected, even if they seem abundant.

If the plants are considered too rare to collect, the best solution is to take several pictures of the plant. Get up as close as possible to the plant and take photographs from several angles and of different parts of the plant.

Ferns & Allies

The ferns, like other groups of plants, are associated with some specialized terminology. The green "leafy" structures are called fronds. The frond is composed of a central stem which is divided into the stipe, below the "leafy" portion, and the rachis, the central vein in the green "leafy" portion. The "leafy" portion itself is the blade. Fronds may occur singly or they may occur in clumps.

The blade may be entire (without lobes or teeth) or, as it is in all of the ones we will see, be divided and subdivided into leaflets and subleaflets. The extent of subdividing is an important characteristic for determining species. Different books use different words to describe the number of times the blade is divided. Perhaps the easiest system is to refer to the various levels of division as "cuts". A blade that is 1-x cut is divided in lobes that reach down to the central vein, the rachis. In a blade that is 2-x cut each of those lobes is in turn divided into lobes that reach down to the central vein of the lobe. Similarly, in a blade that is 3-x cut the lobes of the 2-x cut blade are again divided by lobes that reach down to the central vein of the lobe.

Each level of cut can be modified, so that one can refer to a blade that is ,for example, "almost 1-x cut" where the lobes do not reach quite as far down as the rachis. Or "2-x plus cut" where the third level of cutting is only somewhat completed. The sporophytic generation of ferns, the generation we see, reproduces by asexual spores produced in structures called sporangia. Often these sporangia are clumped together in a sorus (plural sori); in some species the sorus is covered or partly covered by a flap of tissue called the indusium (plural indusia).

Equisetum - Scouring Rushes and Horsetails:

The jointed nature of the stems readily distinguish members of the genus *Equisetum* from other plants. There are a few structures you need to know to understand the descriptions. Ridges extend longitudinally up the stems (diagram below left) as far as the node. The node is where the sections of the stem can be pulled apart (another common name for the group is Jointweed). At the node the ridges extend upward and become the tiny leaves: at first the leaves are fused together and form the leaf sheath, then they separate and become the teeth. The color and shape of the sheaths and teeth are used to determine the species of *Equisetum*. The teeth may fall off during the growing season or they may persist. In some species the teeth break off along a neat line (the articulation line) that is visible before they break away. In most species the stem contains a large central canal and smaller canals along the periphery. In some cases the shape of the reproductive cone (diagram below right) is used; the cone below shows a pointed tip.

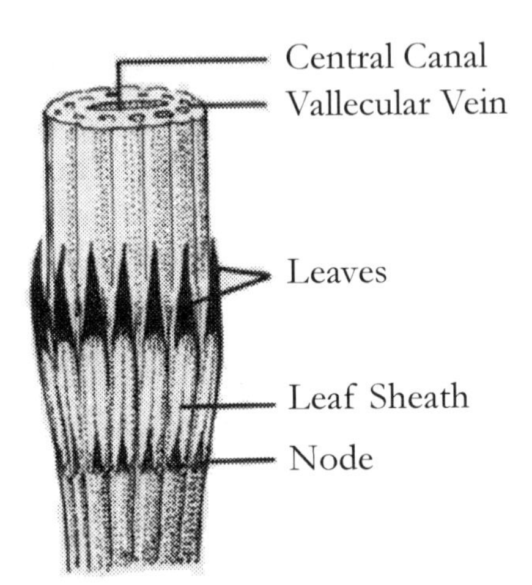

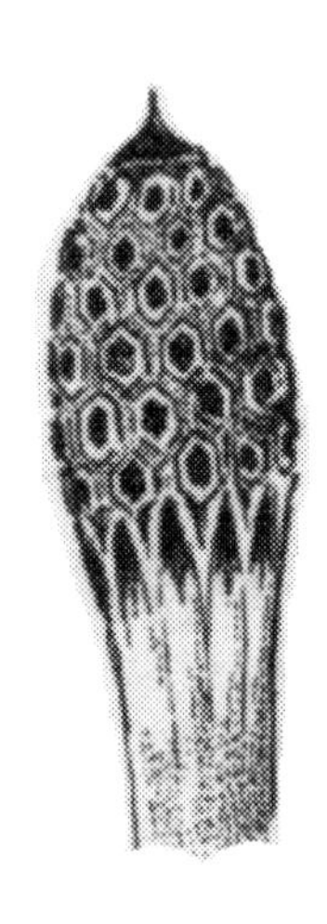

Quillwort

Isoetes spp L.

Status: OBL
Isoetaceae

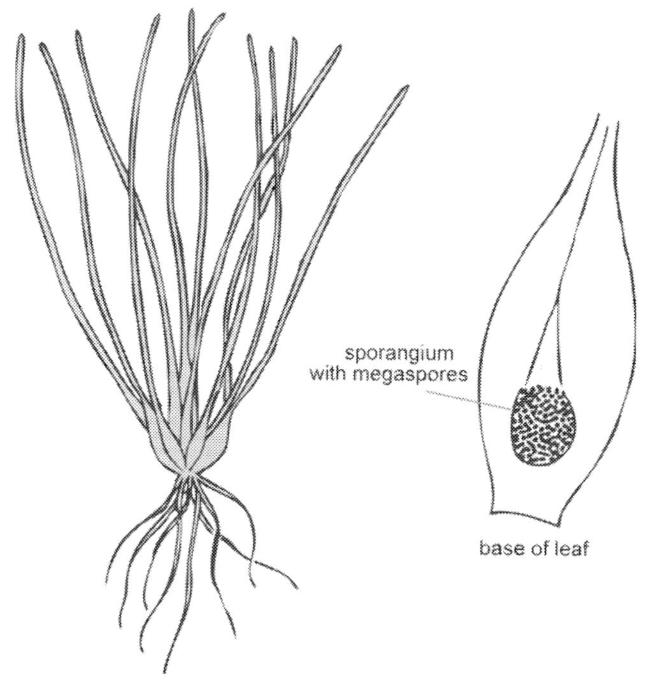

Plants: Usually submerged; looks like a small bunches of chives, seldom more than 30 cm high.
Leaves: Unbranched; rounded or with slight angles; bases of leaves are spoon-shaped and grow in a rosette around a central corm.
Reproduction: The spore bearing sporangia are found on the insides of the bases of the middle and outer leaves of the rosette. During summer each contains either the small microspores or the large megaspores.
Habitat: Shallow waters of lakes and ponds.
Notes: Identification to the individual species level depends on the microscopic examination of the megaspores and is beyond the scope of this book. Six species have been found in the Adirondacks: Appalachian Quillwort, *Isoetes engelmanii* ; Lake Quillwort, *Isoetes lacustris* ; Shore Quillwort, *Isoetes riparia* ; Spiny Spored Quillwort, *Isoetes tenella* ; Tuckerman's Quillwort, *Isoetes tuckermanii*; and an unnamed hybrid, *Isoetes x eatonii.*

Field Horsetail
Equisetum arvense L.

Status: FAC
Equisetaceae

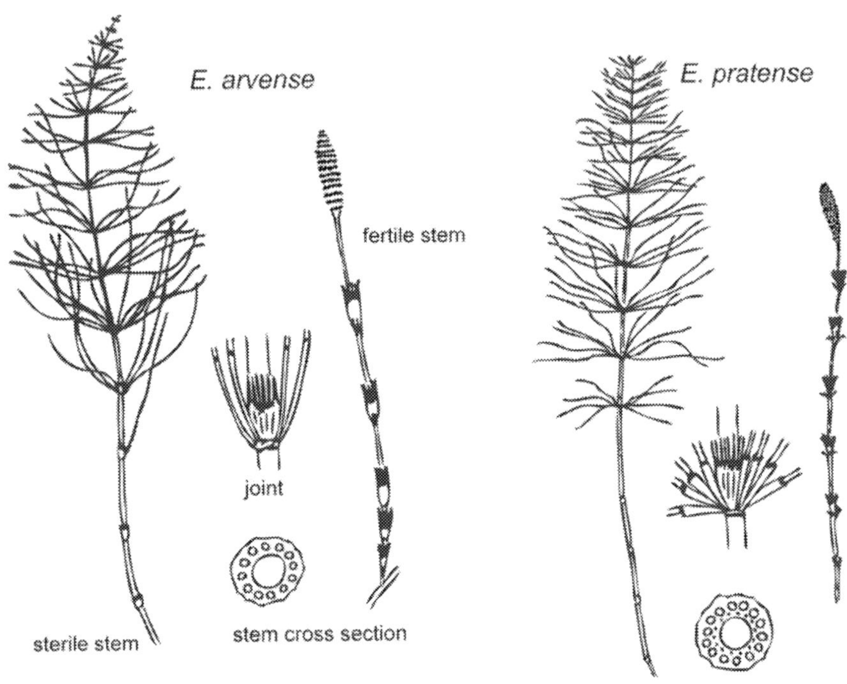

Sterile Stems: to 50 cm, mostly erect but sometimes partly decumbent; rough surface, 12 or more ridges
Nodes: ~ 5 cm apart; sheaths widening upward, about as tall as wide, light brown at base, dark brown at top, topped by 3-4 sharp-tipped black teeth
Branches: horizontal or up-swept, solid, rough; lowest branch usually as long or longer than upper branches; 3-4 angled; length of first node from stem as long as or longer than the sheath from which it arises
Cone: 2-3 cm long, on long stem, blunt-tipped
Habitat: grows in wide variety of habitats, from dry sand banks to moist seme-shaded woodlands
Similar Species: most likely to be confused with Meadow Horsetail, *E. pratense*, in which the length of the first node from the stem is shorter than or just equal to the sheath from which it arises

Water Horsetail

Equisetum fluviatile L.

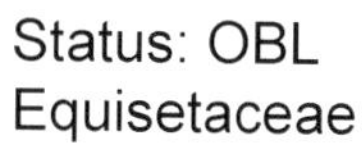

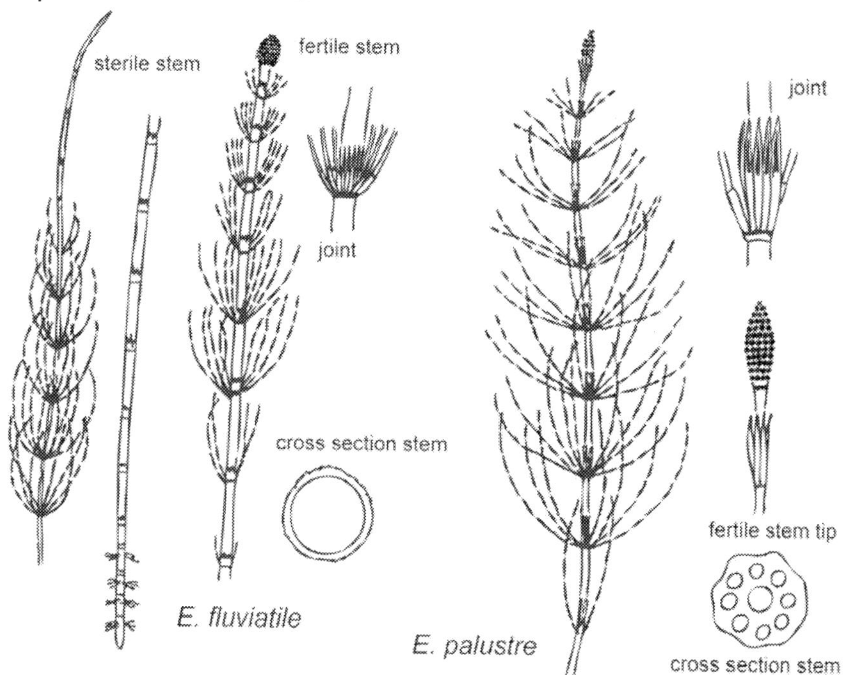

Sterile Stem: to 1 m; very variable with many, few, or no branches; smooth, ~ 20 ridges

Nodes: ~ 5 cm apart; sheaths green, clasping stem tightly; teeth narrow, sharp-pointed, dark, pressed firmly against stem

Branches: if present, usually in the middle third of the stem; smooth, whorled, hollow, variable in length and number; length of first internode less than or equal to the length of the sheath of the node from which it arises

Cavity: central cavity very large, ~ 4/5 diameter of stem; vallecular canals usually lacking

Fertile Stems: similar to sterile stems but unbranched until after the cone forms

Cone: 2-3 cm, blunt-tipped, short-stemmed

Rootstalk: smilar to the stem in being hollow

Habitat: usually grows in standing water

Similar Species: Marsh Horsetail, *E. palustre*, has a small central cavity and conspicuous vallecular canals

Scouring Rush

Equisetum hyemale L.

Status: FACW
Equisetaceae

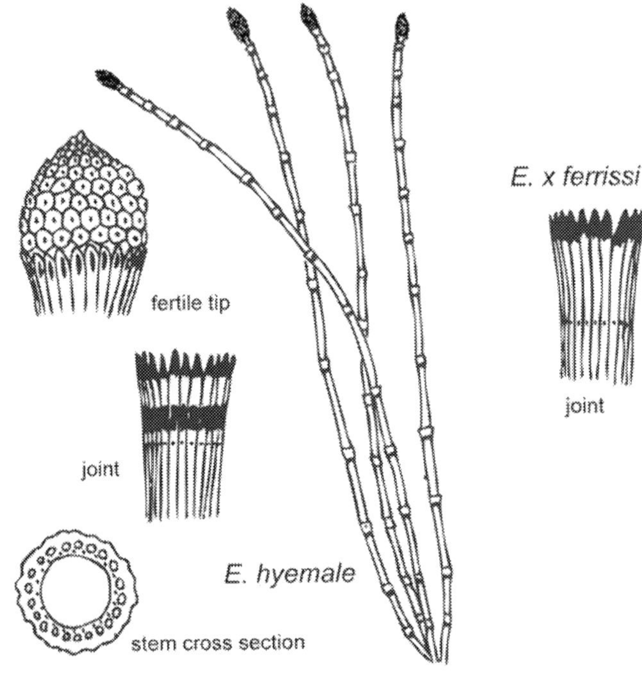

Stems: to 1 m, upright; diameter to 12 mm; rough surface, ~ 30 ridges; fertile and sterile stems alike
Nodes: to 10 cm apart; sheaths about as long as wide, tightly adhering to stem; sheaths green when young, then ageing to gray with a dark girdle at the top, just below the teeth, and another near the base; teeth dark-centered with white margins but quickly wither and fall off along a distinct articulation line
Cavity: central cavity large, ~ 2/3 stem diameter; vallecular canals small
Cone: short-stemmed, sharp-tipped
Habitat: moist soils of roadsides, woodlands, lakeshores
Similar Species: much larger than Dwarf Scouring Rush, *E. scirpoides*, or Variegated Scouring Rush, *E. variegatum*; it is very similar to the hybrid *E. x ferrissi*, but in *E. hyemale* the sheath is about as long as wide and the sheaths of the nodes have a second dark girdle at the base of the sheath

Woodland Horsetail

Equisetum sylvaticum Michx.

Status: FACW
Equisetaceae

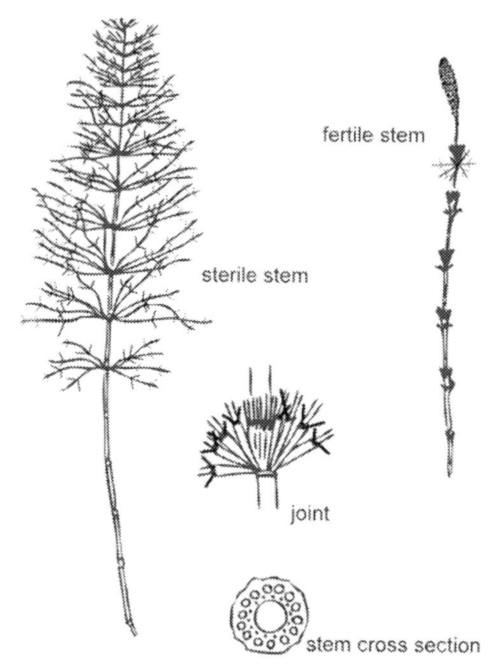

Sterile Stems: to 45 cm, mostly erect, slightly roughened surface, ~ 12 ridges
Nodes: ~ 4 cm apart; green at base, chestnut brown at top; teeth long, narrow, triangular, persistent, often adhering to one another in 3-4 groups
Branches: horizontal to drooping; lowest branch usually shorter than upper branches; 4-angled; branches divide into smaller 3-angled branchlets
Cavity: central cavity large, ~ 1/2 diameter of the stem; vallecular canals large
Fertile Stem: to 20 cm, erect, at first straw-colored, sometimes with small vestigal branches, then (after cone disappears) becomes green and true branches appear, appearing much like sterile stem
Cone: 2-3 cm long, on long stem, blunt-tipped
Habitat: moist forests
Similar Species: the only *Equisetum* where the branches divide into branchlets

Variegated Scouring Rush

Equisetum variegatum Schleich.

Status: FACW
Equisetaceae

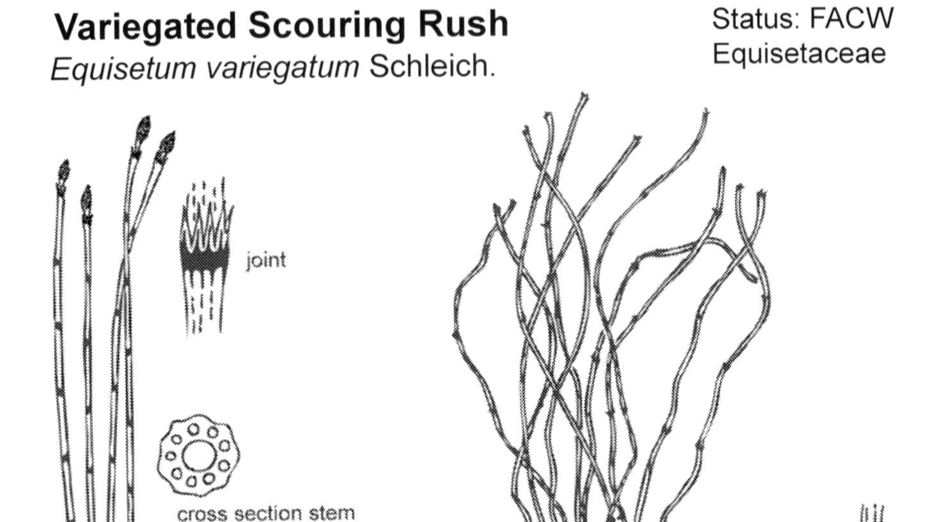

Stems: to 30 cm; lower portion may be decumbent, but upper portion erect; dark green, ~ 3 mm diameter, 5-10 ridges; sterile and fertile stems alike
Nodes: 2-3 cm apart; sheaths funnel-shaped, upper portion black; teeth prominent, sharp-pointed, black center with distinct white edges
Cavity: central cavity ~ 1/3 stem diameter; vallecular canals large; carinal canals small
Cone: short-stemmed, sharp-pointed tip; diameter larger than stem
Habitat: moist woods, beaches, meadows, streambanks
Similar Species: larger and straighter stemmed than Dwarf Scouring Rush, *E. scirpoides*; smaller than the Scouring Rush, *E. hyemale* or *E. x ferrissi*

Lady Fern

Athyrium filix-femina (L.) Roth

Status: FAC
Dryopteridaceae

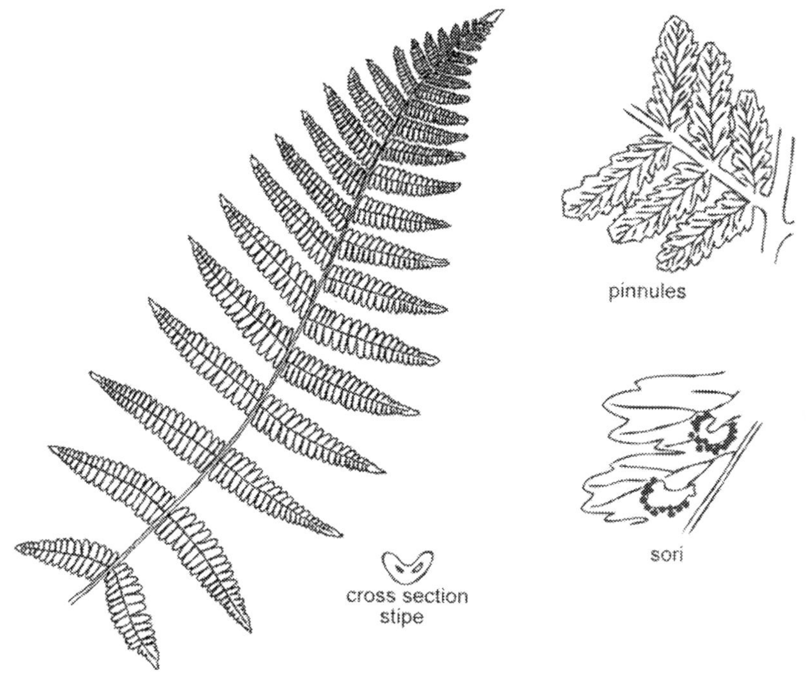

Fronds: 30 to 100 cm; stipe about 1/2 of that
Stipe: green, deeply furrowed in x-section; with a few blackish scales at base
Blade: 2-x plus cut, sometimes almost 3x; semi-tapering; rachis green and furrowed in x-section
Sproangia: in elongate sori, usually comma-shaped and crossing over a vein in the pinnule; indusium a flap-like covering arising from the basal side (may be absent or invisible); sori in rows on each side of central vein, more or less centrally located between the margin and the center
Habitat: moist woods
Similar Species: the lacy 2-x plus cut blade with semi-tapering shape and the comma-shaped sori distinguish this species

Crested Wood-Fern

Dryoptereis cristata (L.) Gray

Status: FACW+
Dryopteridaceae

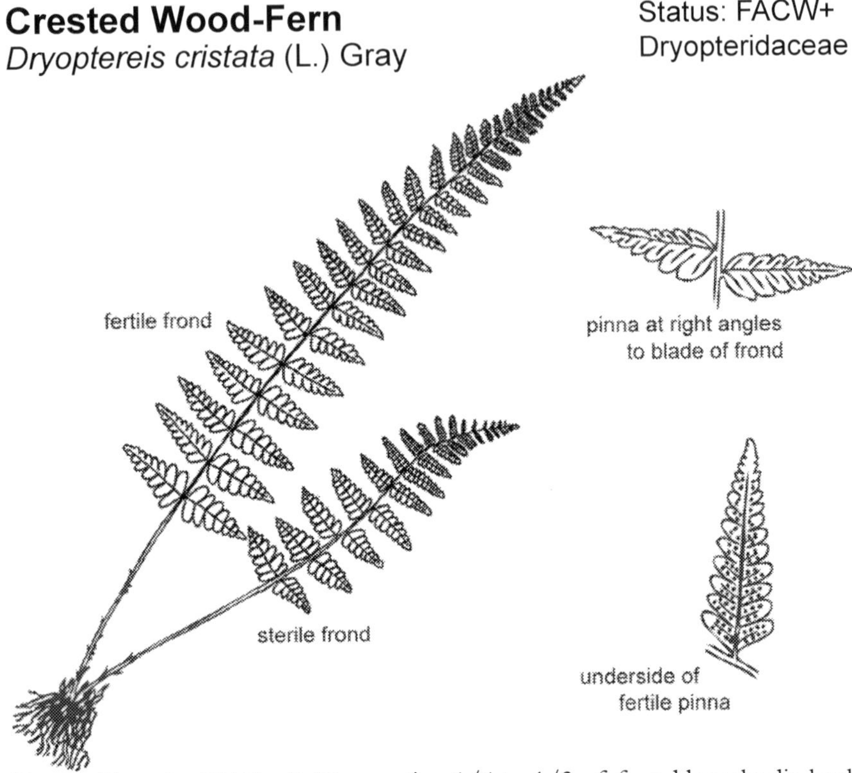

Fertile Fronds: 35-70 x 8-12 cm; stipe 1/4 to 1/3 of frond length; die back during winter
Stipe: tan, scattered scales at the base
Blade: narrowly oblong; 2x to 2x plus cut; pinnae, at least basal ones, broadest near stipe, more or less triangular; in this fern, unlike others, the pinnae are twisted at right angles to be plane of the blade so that they are horizontal
Sporangia: in rounded sori covered by kidney-shaped indusia; midway between centers and margins of secondary pinnae
Sterile Fronds: noticable narrower and often shorter than sterile fronds; several small sterile fronds will form a basal rosette which overwinters
Habitat: marshes, bogs, swamps, shrubby wetlands
Range: see map
Similar Species: easily confused with Clinton's Wood-Fern, *Dryopteris clintoniana*; but the twisting of the pinnae of the fertile fronds so that they are horizontal distinguishes *D. cristata* from other species of *Dryopteris*

Ostrich Fern

Matteuccia struthiopteris (L.) Todaro

Status: FACW
Dryopteridaceae

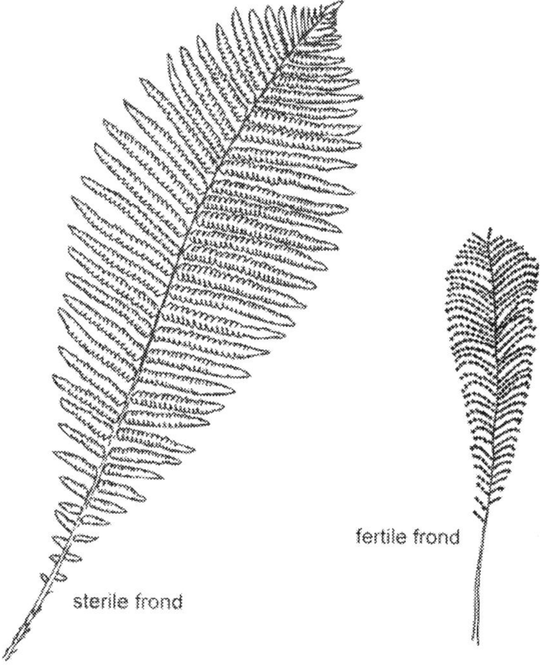

Sterile Fronds: to ~ 2 m; stipe much shorter than blade, nearly lacking; fronds clumped
Stipe: stout, rigid, deeply grooved in front
Blade: to ~ 35 cm wide; oblong lance-shaped; widest above the middle tapering gradually to the base; almost 2-x cut; leaflets alternate; subleaflets oblong, round-tipped, veins not forked; rachis green
Fertile Frond: to ~ 60 cm; stiff, green then brown; 1-x cut; leaflets contract around sori to form long pods
Sproangia: in round sori
Habitat: along streams and riverbanks, in swamps and floodplains; moist neutral to alkaline soil, partial sunshine
Similar Species: Sensitive Fern, *Onoclea sensibilis*, has a somewhat similar fertile frond but the sterile frond is very different

Sensitive Fern
Onoclea sensibilis L.

Status: FACW
Dryopteridaceae

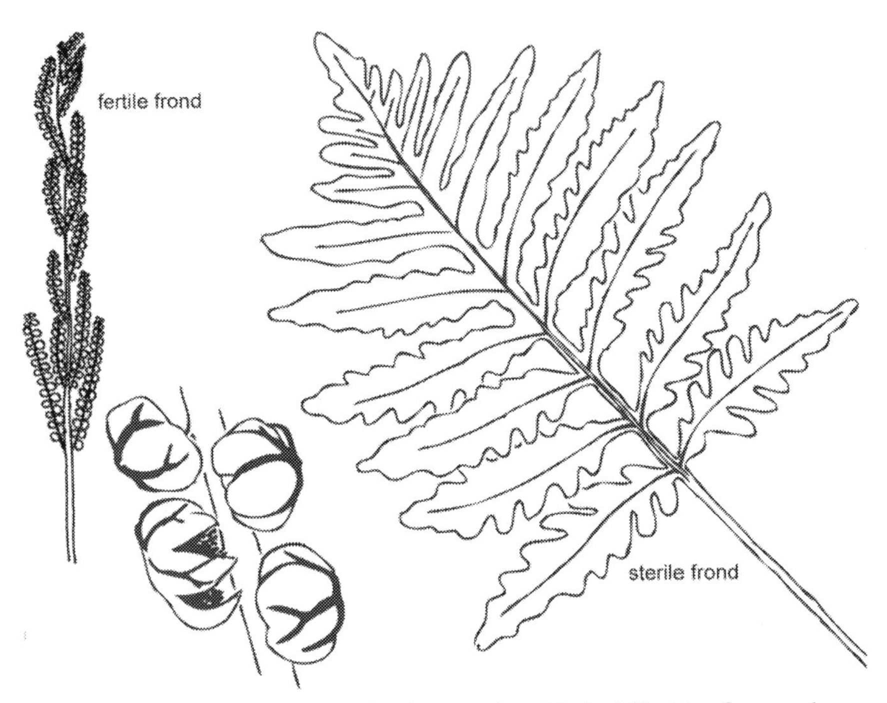

Sterile Fronds: 30 to 75 cm; stipe longer than blade; killed by first cool weather in autumn
Stipe: yellow-brown above to brown at base; shallow groove on front
Blade: almost 1-x to 1-x cut (usually lower part of the blade cut completely to rachis while upper part is not completely cut); leathery, light-green, scattered white hairs on the underside; prominent net-forming veins visible on undersides; rachis yellow-brown, smooth
Fertile Fronds: to 30 cm tall, upright, numerous short lateral branches; sori within small, curled, bead-like segments of the leaflets
Habitat: moist or wet places, shade or full sun
Similar Species: Christmas Fern, *Polystichium acrostichoides*, another 1-x cut fern, has a much longer blade, is evergreen, and does not have separate fertile fronds.

Cinnamon Fern

Osmunda cinnamomea L.

Status: FACW
Osmundaceae

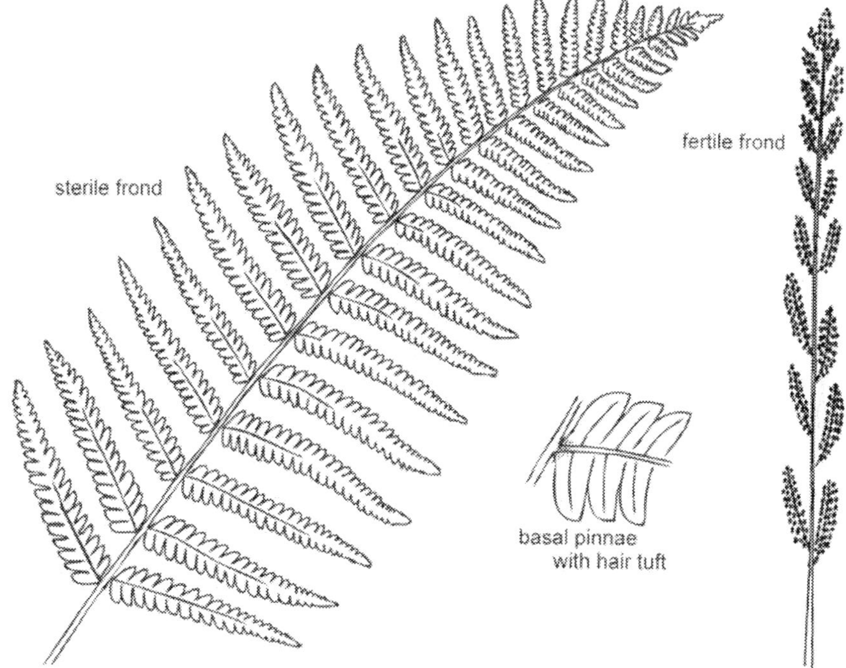

Fronds: to 1 m, erect and arching; stipe 1/4 to 1/3 of length; grows in clumps from a heavy rootstock covered with matted hairs
Stipe: round with semi-grooved face; green; smooth, covered with cinnamon wooly hairs early in season
Blade: 15-25 cm wide; lanceolate; 2-x cut opposite leaflets; dense tuft of rusty-wooly hairs where vein of primary leaflet joins the rachis; rachis semi-grooved in front; smooth with cinnamon-colored wooly hairs early in season
Fertile Fronds: appear before the sterile fronds and wither early in the summer; 1-x cut; leaflets growing upwards hugging the main stem; bright green then turns bright cinnamon-brown
Sporangia: spores in clusters of short-stalked cases
Habitat: damp or waterlogged shady locations
Similar Species: can be distinguished from the fronds of Interrupted Fern, *O. claytonia*, lacking fertile leaflets by the tuft of rusty-wooly hairs at the base of the leaflet

Interrupted Fern

Osmunda claytoniana L.

Status: FAC
Osmundaceae

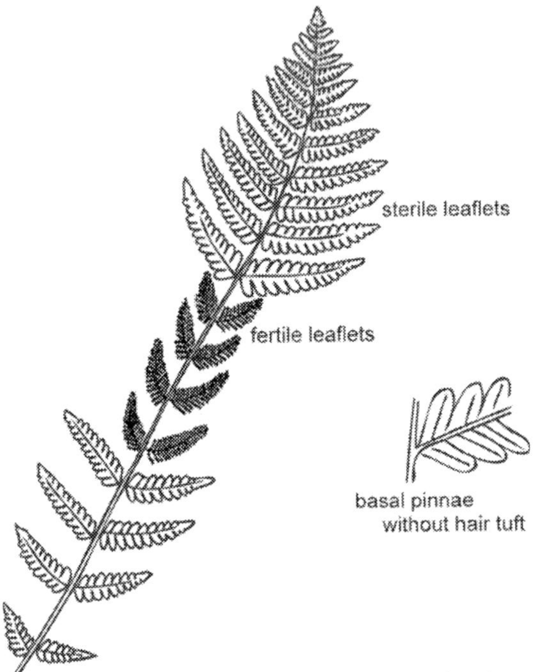

Fronds: 0.7 to 1.0 m; stipe ~ 1/3 of length; grows in clumps; stipe of fronds with fertile leaflets much longer and more upright than others
Stipe: round with semi-grooved face; green
Blade: 15-30 cm wide; oblong-oval; broadest at the middle; wooly at beginning of season, then hairless; 2-x cut; leaflets opposite
Fertile Leaflets: 4 or more central leaflets bear spores, wither and drop off in midsummer
Sporangia: dense cylindrical clusters of short-stalked cases
Habitat: moist to dry woods, semi-shaded areas
Similar Species: easily identified if fertile leaflets present (or after withered away); if not present, as in young clumps, can be distinguished from Cinnamon Fern, *O. cinnamomea*, which has a tuft of rusty-wooly hairs at the bases of the leaflets that is lacking in Interrupted Fern.

Royal Fern

Osmunda regalis L.

Status: OBL
Osmundaceae

Fronds: to 2 m; stipe about 1/3 of length; grows in clumps from an elongated rootstock
Stipe: almost round, slightly grooved in front; straw colored above, reddish at base
Blade: oblong-ovate; 2-x cut; leaflets opposite, widely spaced; fertile leaflets at end of stalk; subleaflets narrow-oblong on small distinct stems, alternate; rachis slender, round, green or straw-colored
Fertile Leaflets: at end of sterile portion of frond; without vegetative tissue, brown
Sporangia: spores in clusters of short-stalked cases
Habitat: wetlands, along streams, in bogs; can grow in standing water
Similar Species: widely spaced leaves and fertile leaflets at the end of the frond are very distinctive

Marsh Fern

Thelypteris palustris Schott

Status: Unknown
Thelypteridaceae

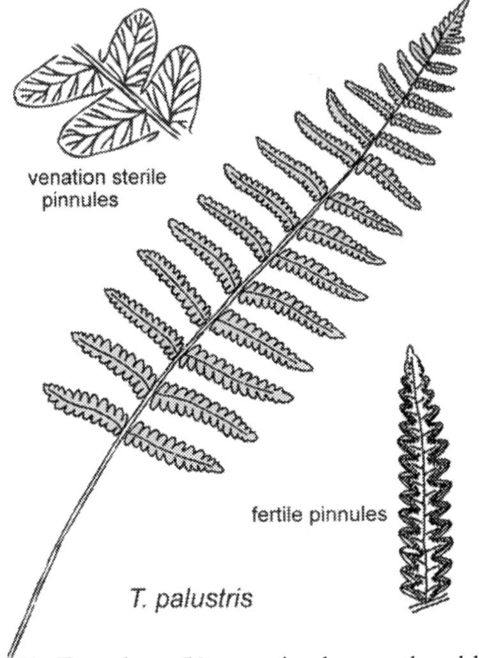

venation sterile pinnules

T. simulata

fertile pinnules

Sterile Fronds: ~ 50 cm; stipe longer than blade
Stipe: stipe smooth; pale green above, black at base
Blade: ~ 15 cm wide; lance-shaped, widest just above base; almost 2-x cut; lowest pair of leaflets perpendicular to rachis and at least 1/2 as long as longest leaflets; at least some of the veins of leaflets forked between midvein and margin of leaflet; green to yellowish-green, thin & delicate; rachis green, slender, smooth
Fertile Fronds: more erect than sterile with longer stipes; margins of leaflets curled over sporangia
Habitat: open woodlands, moist meadows, shores, and ditches; sunny or partial shade
Similar Species: the inrolled margins of the leaflets on the fertile fronds distinguish this species from most others of comparable blade shape and "cutting"; in the similar Massachusetts Fern, *T. simulata*, the veins in the leaflets are not forked

Woody Plants

In these species the leaves are attached to a woody stem which persists from year to year and typically adds an increment of growth each year. In the Adirondack region this is a fairly discrete group of plants and there are only a few species where the distinction is not obvious. For the identification of the woody plants, a very simple key is constructed as follows:

1a) Leaves needle-like or scale-like … Conifers *(page 24)*
1b) Leaves broad and deciduous … 2

2a) Leaves opposite or whorled … 3
2b) Leaves alternate … 4

3a) Leaves compound … Opposite Compound Leaves *(page 28)*
3b) Leaves simple … Opposite Simple Leaves *(page 32)*

4a) Leaves compound … Alternate Compound Leaves *(page 46)*
4b) Leaves simple … 5

5a) Leaves fan-veined … Alternate Simple Leaves: Fan-veined *(page 55)*
5b) Leaves ladder-veined … 6

6a) Leaves entire … Alternate Simple Leaves: Ladder-veined; Entire * *(page 61)*
6b) Leaves toothed or lobed … Alternate Simple Leaves: Ladder-veined; Toothed or Lobed *(page 72)*

*: Some willows have entire leaves, others toothed. *(Go to page 91)* for further information on willow identification.

Balsam Fir

Abies balsamea (L.) P. Mill.

Status: FAC
Pinaceae

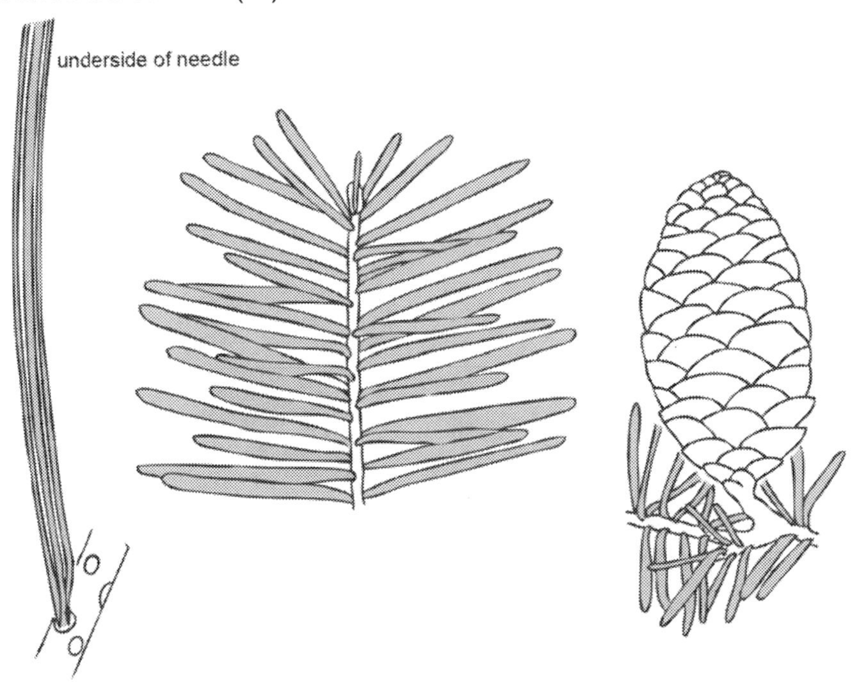

Needles: 10-22 mm long, flattened with blunt tips; 2 white stripes paralleling the midvein underneath; needles attached directly to twigs on a broad circular base; twigs smooth where needles are missing
Cones: 2.5-7.0 cm long; stand upright on twigs; scales tightly packed, cone solid; young cones purplish to greenish, becoming brown when mature.
Trees: up to 15 m tall, dimaeter up to 45 cm; bark smooth with hortizontal resin blisters
Habitat: moist woods and swamps.
Similar Species: the only other conifer with flattened blunt needles that are whitened beneath is the Hemlock, *Tsuga canadensis*, in which the needles have stalks and are attached to small projections on the twig; the difference in the bark is also a good character to distinguish between Balsam Fir and Hemlock

Tamarack

Larix laricina (DuRoi) K.Koch

Status: FACW
Pinaceae

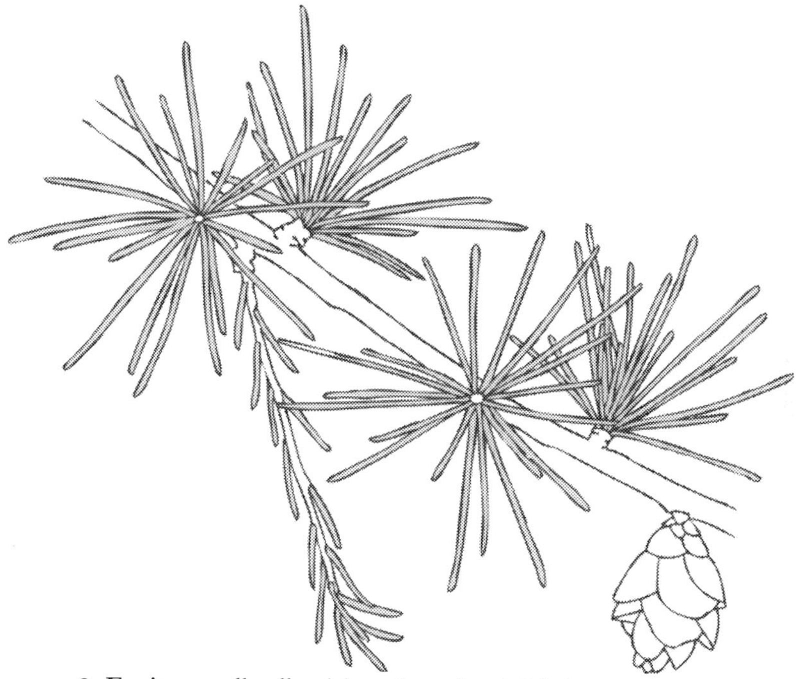

Flowers & Fruits: small yellowish male and reddish-brown female cones, in clusters on dwarf branches of previous year's growth
Leaves: deciduous; flat needles, 2-5 cm long, 3-5 mm wide, light to bluish green; either in clusters of 15-60 on short lateral branches from previous year's growth or scattered along shoots of current year's growth
Twigs: light brown to orange-brown, slender, short
Bark: gray to reddish brown to brown, scaly, smooth as a young tree and rough when aged
Tree: up to 20 m tall; diameter to 60 cm
Habitat: bogs, swamps; but grows best on moist well-drained soils

Black Spruce

Status: FACW-
Pinaceae

Picea mariana (Mill.)
Britton, Sterns, & Poggenb.

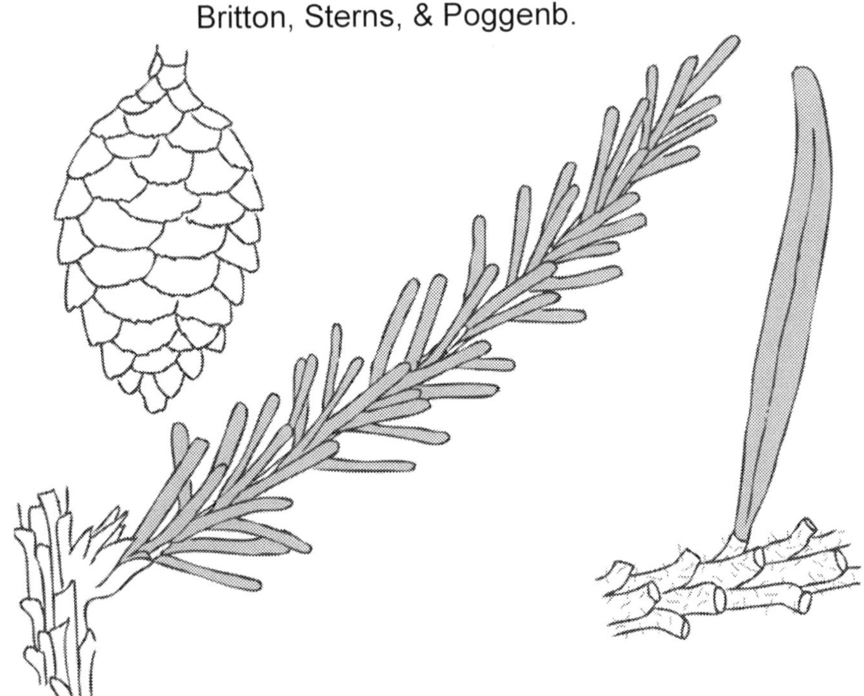

Cones: pistillate cones dark purple before maturity, point downward, dull gray-brown when ripe; 1.5-3.5 cm; persist many years; scales with thin brittle wings

Needles: squarish, 4-sided, stiff, ends often blunt; arise on small pegs from appressed branchlets; 6-18 mm; dark grayish-green

Buds & Twigs: twigs covered with grayish crooked hairs (use hand lens); buds gray-brown buds with loose scales

Bark: gray to brown, close scales

Tree: to 10 m, occasionally larger; narrow spire-like crown, branches tend to droop

Habitat: sphagnum bogs, acidic swamps; in the Adirondacks Red Spruce, *P. rubens*, also often found in wetland habitats

Similar Species: Red Spruce, *P. rubens*, has twigs with reddish crooked hairs and shiny yellow-green needles; White Spruce, *P. glauca*, has twigs without hairs, cone scales with flexible margins, and green needles; in both species the tree has a more pyramidal shape

Northern White Cedar
Thuja occidentalis L.

Status: FACW
Cupressaceae

Leaves: scale-like, hugging the twigs and branchlets; about 2-4 mm long. Arranged in 4 rows around twig, but strongly flattened. Twigs in flattened sprays
Cones: about 10-15 mm long, brown, with loose scales when ripe.
Bark: fibrous, long vertical ridges which may tend to spiral around the trunk
Trees: to 15 m tall; diameter to 60 cm
Habitat: moist or wet soil, often in swamps
Similar Species: this is the only conifer in the Adirondacks that has only scale-like leaves; most *Juniperus* have three-sided needles but note that Common Juniper, *J. virginianus*, may have scale-like leaves that are not flattened in addition to three sided needles; the Junipers also have a hard resiny berry as a fruit
aka: Arbor Vitae

Box Elder

Acer negundo L.

Status: FAC+
Aceraceae

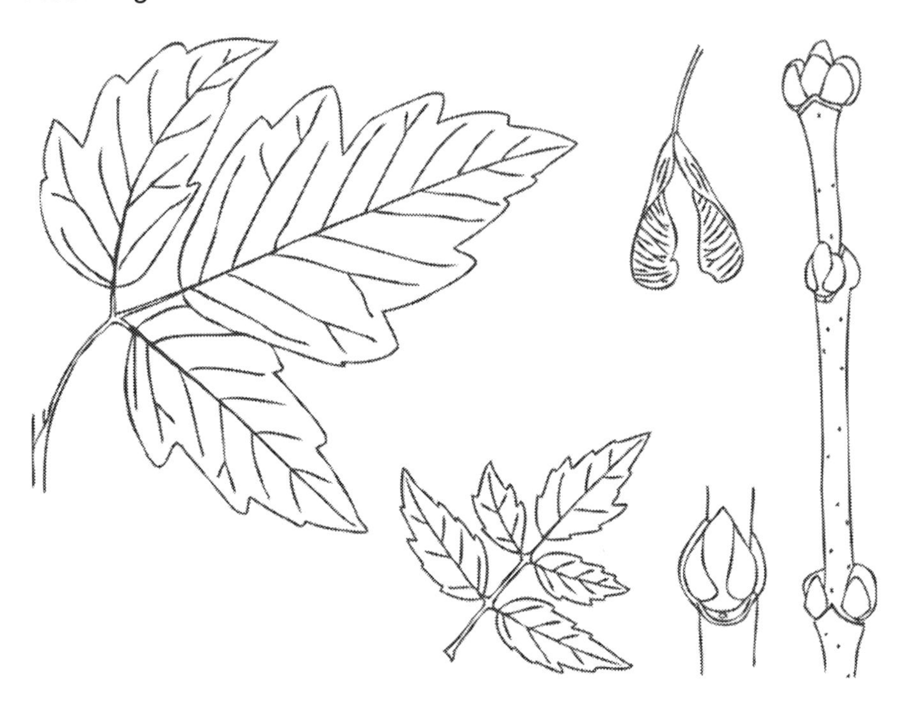

Leaves: 10-25 cm long, 3-5 leaflets (occasionally up to 9); leaflets coarse-toothed to lobed, often assymetrical
Buds: white-hairy
Twigs: purplish or greenish; leaf scars narrow and wrap around the twig to meet on the opposite side; 3 bundle scars
Bark: furrowed
Tree: to 20 m high; diameter to 1 m
Habitat: moist soil, especially along streambanks
Range: NH west to Pacific; south to Guatemala
Similar Species: differs from other *Acer spp* in having compound leaves; if only 3 leaflets present seedlings can resemble *Toxicodendron radicans* (Poison Ivy), but leaf scars and the lack of a skin reaction distinguish between them

Black ash
Fraxinus nigra Marsh.

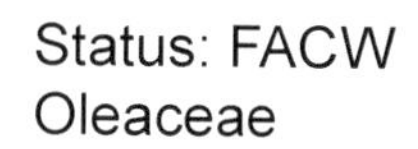

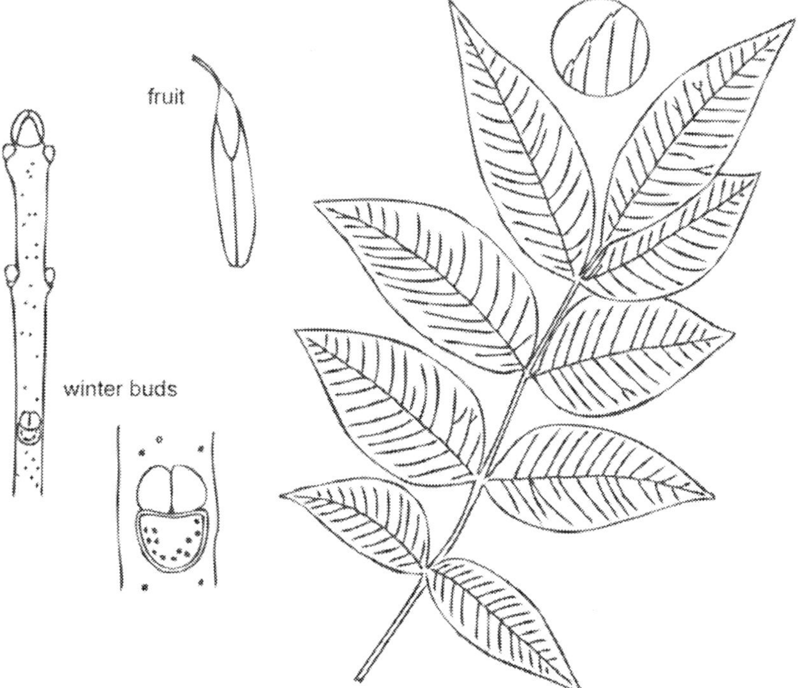

Leaf: 27-40 cm long; 7-11 leaflets, always with sharply pointed teeth; never stalked; in spring has tufts of reddish-brown hairs where the leaflet attaches to leaf-stalk but these are lost dduring the summer
Buds: very dark brown to dark brown (good distinguishing character in winter)
Twigs: grayish brown, dull; leaf scar not deeply notched
Bark: corky and bumpy when young; warty and scaly when older
Tree: up to 20 m tall; diameter to 50 cm
Fruits: blunt on both ends
Habitat: floodplains, wet woods, swamps
Similar Species: can sometimes be difficult to distinguish from the other Ashes, *Fraxinus spp*, as some of the characteristics can overlap - try to use as many characteristics as possible, but the texture of the bark is often one of more useful during the summer

Green ash

Fraxinus pennslyvanica Marsh.

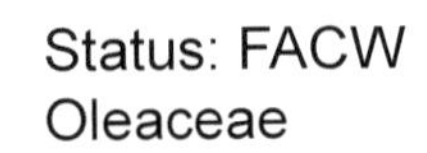

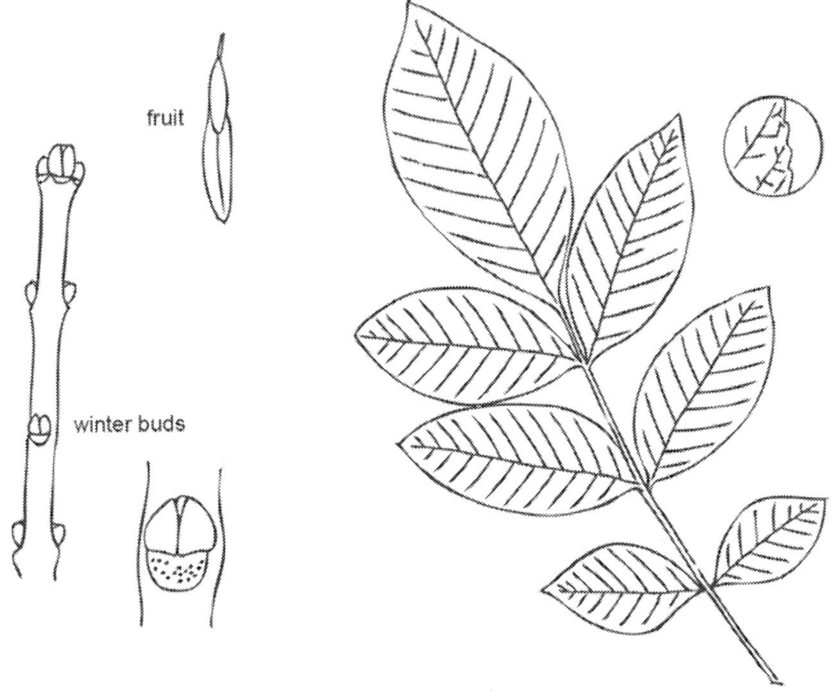

Leaf: 25-30 cm long; leaflets 5-9 (usually 7), usually with dull-teeth, stalked (sometimes very short)
Buds: dark brown to brown
Twigs: dark brown and shiny; leaf scars not notched
Bark: smooth when young, then with well defined but somewhat round-edged ridges
Tree: up to 20 m tall; diameter to 75 cm
Fruit: usually acute at base, blunt at distal end
Habitat: floodplains, wet woods, streambanks
Similar Species: sometimes difficult to distinguish from White Ash, *F. americana* as many of the characteristics can overlap - in White Ash the leaf scar is notched but the depth of the notch varies and many intermediates are present, the ridges in the bark are narrower and have sharper edges, and the leaflets are relatively narrower; try to use as many characteristics as possible

Black Elderberry

Sambucus nigra (L.H. Bailey) Fernald

Status: FACW-

Caprifoliaceae

Flowers & Fruits: flowers small (3-4 mm across), white, 5-parted; in a flat-topped cluster without an obvious main axis; fruits purple, ~ 5 mm in diameter, in a flat-topped cluster

Leaves: compound-opposite, 10-25 cm long; 5-11 leaflets, finely toothed

Buds & Twigs: buds small, green or brown; twigs stout with warty lenticils, pith white; leaf scars large, connected by ridges to opposites, 5-7 bundle scars

Bark: brown

Shrub: to 3 m tall; diameter to 3 cm

Habitat: moist woods, fields, roadsides

Similar Species: Red-Berried Elderberry, *S. racemosa*, is also found in the Adirondacks. It has flowers and fruit clusters that are more raceme-like with a central stalk, the berries are red when mature, and the pith of the twigs is brownish

Red Maple

Acer rubrum L.

Status: FAC
Aceraceae

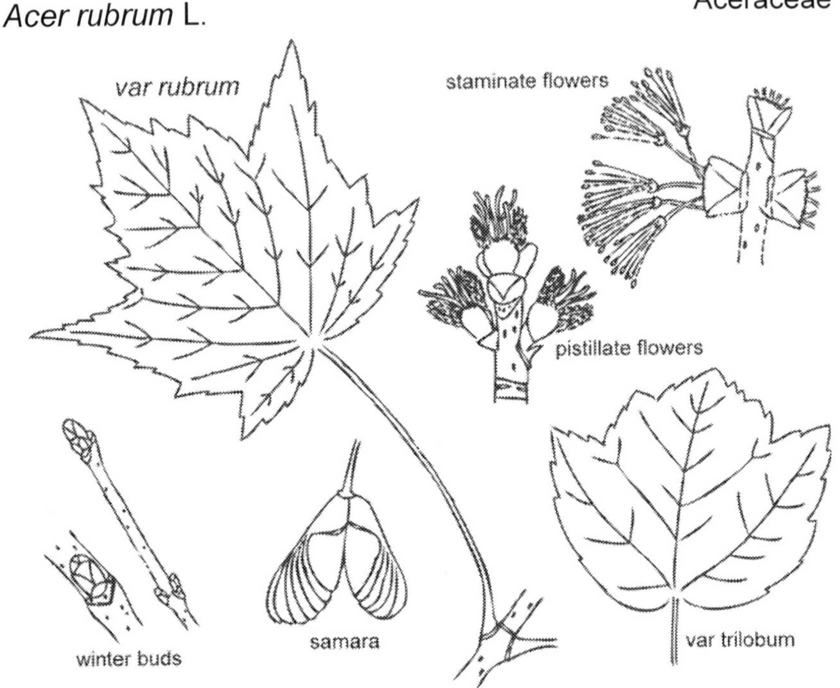

Flowers: red, unisexual, opening well before the leaves
Fruits: samara, 12-25 mm
Leaves: opposite-simple; 5-15 cm long; 5 lobes, 3 major plus 2 minor at base, lobes coarsely toothed; lobes with angles at the base of the notch; terminal lobe about 1/2 or less of the length of the leaf blade, terminal lobe usually does not get significantly narrower near the base; leaves silvery white beneath
Twigs and Buds: reddish, without unpleasant odor when broken
Bark: smooth and gray on smaller trees, rough and dark on older trees
Habitat: swamps, alluvial soil, moist uplands
Similar Species: the variety of Red Maple *trilobum* has only 3 lobes on the leaf and has a FACW+ wetland status; Silver Maple, *A. saccharinum,* has much deeper notches so the terminal lobe is well over 1/2 the length of the leaf blade; the terminal lobe is also significantly narrowed at the base

Silver Maple

Acer saccharinum L.

Status: FACW
Aceraceae

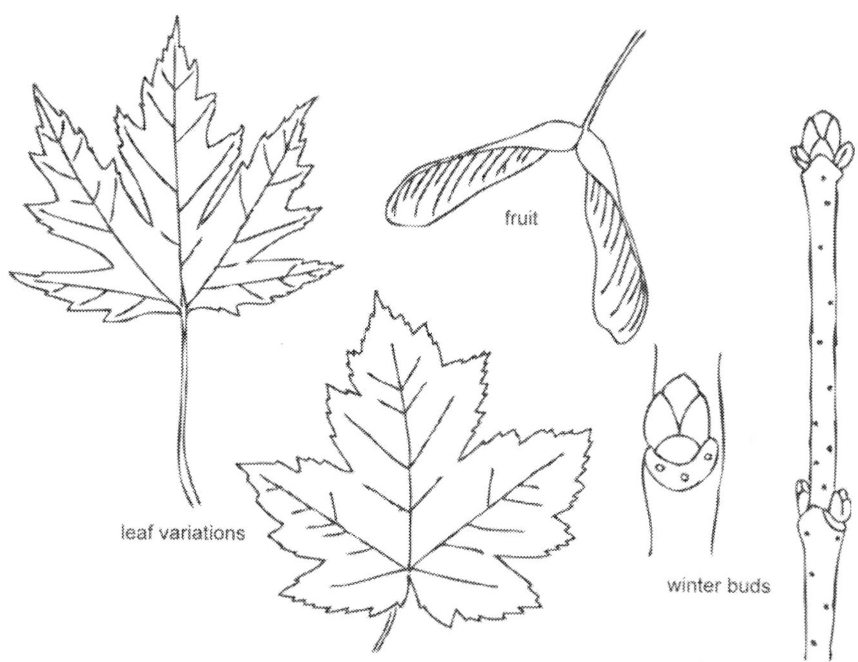

Flower: greenish or reddish, unisexual, opening well before leaves
Fruits: samara, maturing before leaves fully opened
Leaves: opposite-simple, silvery-white beneath; 5 lobes, lobes coarsely toothed; lobes with angles at the base of the notch; terminal notch at least one-half of the length of the leaf blade and narrows significantly toward the base
Twigs and Buds: reddish with an unpleasant odor when broken
Bark: light grayish, flaking in larger plates on older trees
Trees: up to 25 m tall
Habitat: moist and wets soils of riverbanks and floodplains
Similar Species: typical Red Maple, *Acer rubrum,* has shallower notches and the terminal lobe does not narrow significantly toward the base; in Sugar Maple, *Acer saccharum,* the bottoms of the leaf notches are rounded, not angled

Sugar Maple

Acer saccharum Marsh.

Status: FACU-
Aceraceae

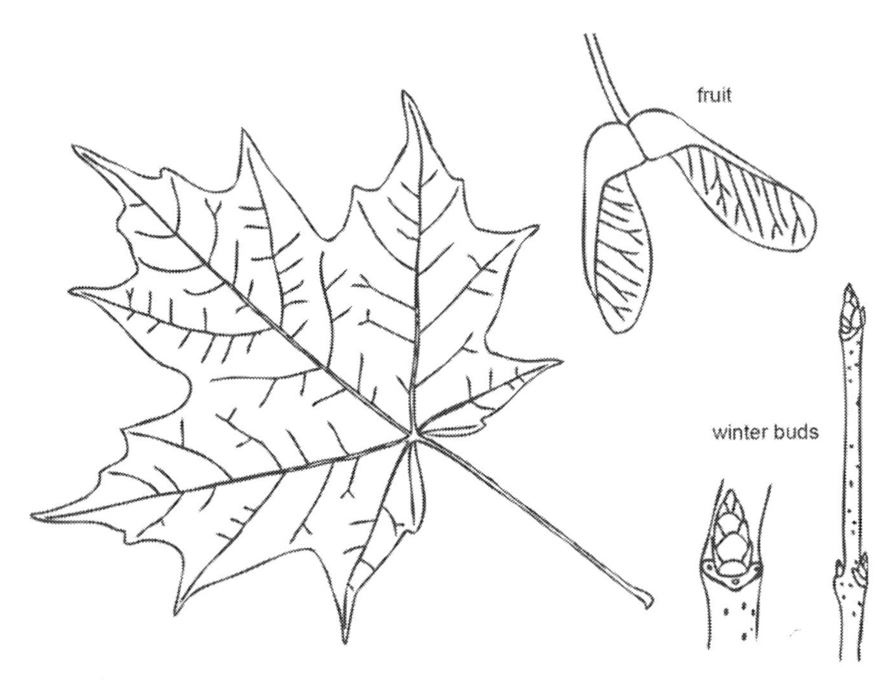

Flowers and Fruits: flowers yellowish, drooping on long (to 8 cm), hairy stalks, opening as leaves open; fruit a samara
Leaves: opposite-simple, green beneath, 5-25 cm long; 5 lobes, each with a few teeth; lobes with bottom of notch rounded
Twigs: glossy, reddish-brown
Bark: dark gray-brown with rough vertical ridges and furrows when mature
Tree: up to 20 m; diameter up to 60 cm
Habitat: mature upland forests
Similar Species: Black Maple, *A. nigrum*, is very similar but leaves have shallower notches, and the edges of the leaf droop and the undersides have a velvety covering of short brown hairs, and the twigs are dull; leaves of Red Maple, *A. rubrum*, and Silver Maple, *A. saccharinum*, have notches with angles at bottom instead of rounded and lobes that are coarsely toothed

Buttonbush

Cephalanthus occidentalis L.

Status: OBL
Rubiaceae

Flowers and Fruits: small (4 mm) white 4-part tubular flowers clustered in dense heads on an elongate stalk; dense heads of small (5mm) fruits remain on the plant through the fall and winter mature fruits
Leaves: opposite or in whorls of 3, simple; 8-15 cm long
Twigs and Buds: lateral buds embedded in the twig bark
Shrub: 1-3 m tall
Habitat: shallow ponds, wet shores
Similar Species: the fruits and/or flowers make this species difficult to confuse with anything else

Silky Dogwood

Cornus amomum P. Mill.

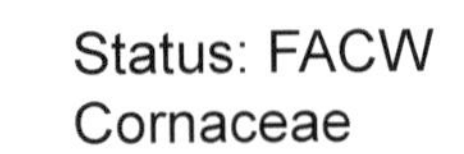
Status: FACW
Cornaceae

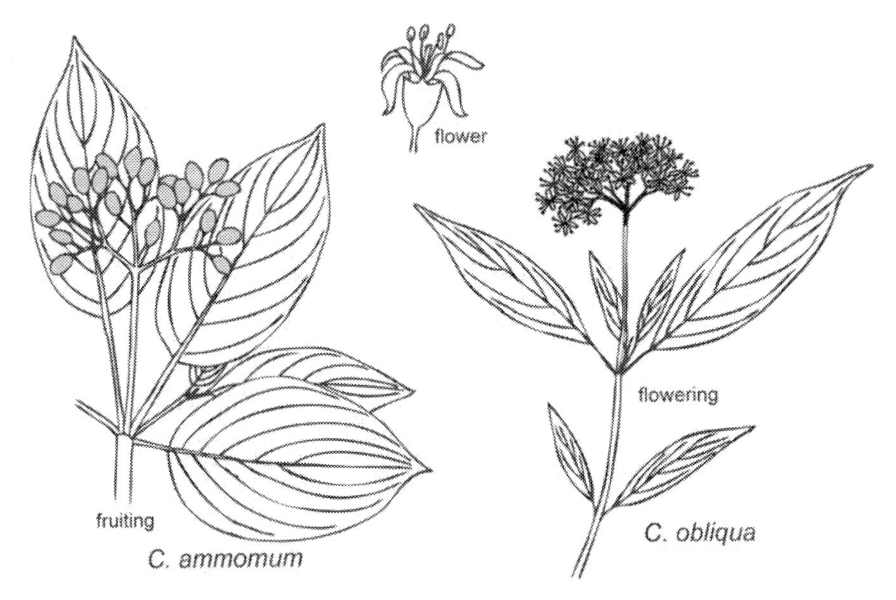

Flowers and Fruits: small (2-4 mm), white, 4-part flowers in open rounded heads; fruits blue

Leaves: opposite-simple; as with all *Cornus*, the branch veins turn and run parallel to the edge of the leaf; leaves 5-10 cm long with rounded bases, the length less than twice the width

Buds and Twigs: twigs and small branches dull purple, silky-hairy when young; pith dark-brown

Shrubs: 1-3 m tall

Habitat: moist or wet soils

Range: ME and Que. west to IL, south to AL, east to FL

Similar Species: among the other *Cornus* shrubs: Narrow-Leaf Dogwood, *C. obliqua,* is very similar but has narrower leaves (length more than twice width) with tapered bases; Alternate-Leaf Dogwood, *C. alternifolia*, has the leaves bunched near the ends of the branches and at least some are alternate; Red Osier Dogwood, *C. sericea,* has redder twigs and white pith; Red-Panicled Dogwood, *C. racemosa,* has gray-brown twigs, red panicles on the fruits, narrower leaves, and light brown pith; Round-Leaf Dogwood, *C. rugosa*, has almost round leaves

Red Osier Dogwood

Cornus sericea L.

Status: FACW+
Cornaceae

Flowers and Fruits: small (2-4 mm), white, 4-part flowers in open rounded heads; fruits white
Leaves: opposite-simple; as with all *Cornus*, the branch veins turn and run parallel to the edge of the leaf; leaves 5-10 cm long, 1.5 to 2.0 times as long as wide, with tapered or rounded bases, whitened beneath
Buds and Twigs: twigs and small branches red to purple; pith white, large
Shrubs: 1-3 m tall
Habitat: streambanks and moist woods
Similar Species: Silky Dogwood, *C. ammomum*, and Narrow-Leafed Dogwood, *C. obliqua*, have brown pith and dull purple branches - (in C. *sericea sericea* and in *C. ammomum*/*C. obliqua* the intensity of the color of the twigs is brightest in earlier spring and dullest in early winter, but at any time of year *C. sericea* is always a brighter, redder color)

Sheep Laurel

Kalmia angustifolia L.

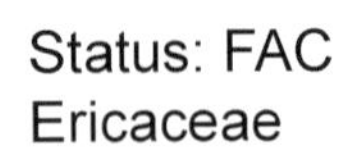
Status: FAC
Ericaceae

Flowers and Fruits: flowers 5-parted, 6-12 mm wide, reddish-purple to pink; fruits hard capsules; flowers and fruits in lateral clusters arising from axils of previous year's leaves, i.e. lateral, not terminal
Leaves: opposite or in whorls of 3, simple; 3-5 cm long, leathery, evergreen, pale green beneath
Twigs: round, hairless
Shrub: to 1 m tall
Habitat: moist open areas, often in bogs, edges of shallow ponds
Similar Species: Bog Laurel, *K. polifolia*, has narrower leaves with the edges rolled under and whitened beneath; its flower clusters are terminal

Bog Laurel

Kalmia polifolia Wangenh.

Status: OBL
Ericaceae

Flowers and Fruits: flowers 5-parted, 10-16 mm wide, rose-purple; fruits hard capsules; flowers and fruits in clusters arising from axils of current year's leaves so flower clusters appear to be at the end of the branch, i.e. terminal, not lateral
Leaves: opposite, simple; 1-4 cm long, leathery, evergreen, edges rolled under (revolute), bottoms whitened with very fine hairs
Twigs: sharply 2-edged, hairless
Shrub: to 1 m tall
Habitat: bogs, edges of shallow ponds
Similar Species: Sheep Laurel, *K. angustifolia* (Sheep Laurel) has wider leaves with flat edges and green beneath; its flower clusters are lateral; the leaves of Bog Rosemary, *Andromeda polifolia*, are similar, though usually somewhat narrower, but are alternate; its flowers, if present, are very different

Swamp Fly Honeysuckle

Lonicera oblongifolia (Goldie) Hook.

Status: OBL
Caprifoliaceae

Flowers and Fruits: bilaterally symmetrical, tubular ending in two lips; paired, 1 cm long, light yellow to reddish; fruit reddish purple to purple berries

Leaves: opposite-simple, ovate, tapering at the base, blunt tips; 3-10 cm long; hairless above and below

Buds and Twigs: pith of newer branches white

Shrub: to 1.5 m

Habitat: bogs, cedar swamps

Similar Species: the introduced Tatarian Honeysuckle, *L. tatarica*, and Morrow's Honeysuckle, *L. morrowii*, both have end branchlets that are either hollow or have brownish pith; the other shrub Honeysuckles in the Adirondacks are American Fly Honeysuckle, *L. canadensis*, and Mountain Fly Honeysuckle, *L. villosa*, both of which have flowers with radial symmetry and leaf tips that are more rounded

Tatarian Honeysuckle

Lonicera tatarica L.

Status: FACU
Caprifoliaceae

Flowers and Fruits: bases of petals fused, forming a tubular corolla with 5 lobes, ~20 mm long; lobes of corolla of almost equal size and symmetrically arranged (radial); corolla pink, sometimes white but not turning yellow with age; fruits red berries
Leaves: opposite, simple; 2.5-5.0 cm long; hairless above and below; leaf stalks 15-25 mm, hairless
Twigs: twigs and small branches hollow; scales remain at the base of the twigs marking the boundary with the previous season's growth; opposing leaf scars connected by lines; bundle scars 3
Shrub: to 2 m
Similar Species: Tatarian Honeysuckle and Morrow's Honeysuckle, *L. morrowii*, both have hollow end branchlets - in the native shrubby species the pith of the branchlets is white; Tatarian Honeysuckle is relatively easy to distinguish because it has leaves that are hairless with longer leafstalks instead of grayish-hairy underneath with shorter leafstalks (5-15 mm); also it has pink flowers pink rather than yellowish white and radial rather than bilateral symmetry

Common Buckthorn

Rhamnus cartharctica L.

Status: UPL
Rhamnaceae

Flowers and Fruits: plants unisexual (male and female flowers in separate plants); flowers 4-parted, greenish white, on stalks in clusters in the axils of leaves; fruits black dull berries 5-6 mm in diameter, in dense clusters, foul taste

Leaves: opposite (occasionally alternate), simple, toothed; 3-6 cm long, no more than twice as long as wide; lateral veins strongly upturned near margins

Buds and Twigs: twigs dark, twig tips usually spiny

Bark: inner bark yellow; outer bark brownish-yellow with horizontal streaks, peels in thin strips

Shrub: shrub or small tree, to 6 m tall

Habitat: hedgerows, old fields, secondary forests

Similar Species: spiny twig tips, berries, and bark distinguish from other opposite-simple leaved woody plants

Southern Arrowwood
Viburnum dentatum L.

Status: FAC
Caprifoliaceae

Flowers and Fruits: flowers grouped in dense rounded head 5-10 cm in diameter; flowers 4-8 mm diameter, 5-lobed, white; fruit blue-black berry, 5-10 mm long; stone deeply grooved on one side
Leaves: opposite-simple, 4-10 cm, sharp coarse teeth (4-22 on each side of leaf), hairy beneath, leaf bases rounded; upper leaf stalks 8-25 mm long
Twigs and Buds: buds several-scaled; twigs hairless with ridges
Bark: gray-brown to reddish
Shrub: 1-5 m tall
Habitat: moist woods and swamps
Similar Species: a threatened subspecies (*venosum*) of Southern Arrowwood has hairy leaf stalks; Southern Arrowwood is also similar to Shortstalk Arrowwood, *V. rafinesquianum*, which has twigs that are not ridged and upper leaves on the branch that have petioles less than 5 mm long

Wild Raisin

Status: FACW
Caprifoliaceae

Viburnum nudum L. *var cassinoides* (L.) Torr & A. Gray

Flowers and Fruits: Flowers 5-lobed, 4-8 mm diameter, white; grouped in dense rounded head 5-10 cm in diameter, head without a central stalk; fruits blue-black, 6-12 mm long with a sweet pulp; stone flattened, not grooved
Leaves: opposite-simple, sharply toothed, green, 5-12 cm long; stalks with "wings" along the side; leaves without long pointed tips
Twigs and Buds: 2-scaled with elongate tips
Shrubs: to 10 m tall
Habitat: wet woods and swamps
Similar Species: very similar to Nannyberry, *V. lentago*, in which some of the leaves always have elongate tips and the flower/fruit cluster has an obvious central stalk

Cranberry Viburnum

Viburnum opulus L.

Status: FACW
Caprifoliaceae

Flowers and Fruits: flowers grouped in dense rounded head 5-10 cm in diameter; outer flowers much larger (15-25 mm diameter) than inner (4-8 mm); flowers 5-lobed, white; fruit a red berry, 10-15 mm long; stone flattened, smooth

Leaves: opposite-simple, 3 lobes with 3-5 veins meeting at the base; sharply toothed, green to reddish-green, 5-10 cm long, hairy beneath, especially on the veins; 1-6 glands on leaf stalk near base of leaf

Twigs and Buds: buds 2-scaled; twigs smooth

Shrubs: to 5 m

Habitat: moist woods

Similar Species: leaves similar to Squashberry, *V. edule*, and Mapleleaf Viburnum, *V. acerifolium*, both of which lack the larger outer flowers; when no flowers are present Mapleleaf Viburnum can be distinguished by its hairy twigs, Squashberry by the lack of glands on the leaf stalk near the base of the leaf; Hobblebush *V. alnifolium* has similar flowers to Cranberry Viburnum, but the leaves are not lobed

Butternut

Juglans cinerea L.

Status:FACU-
Juglandaceae

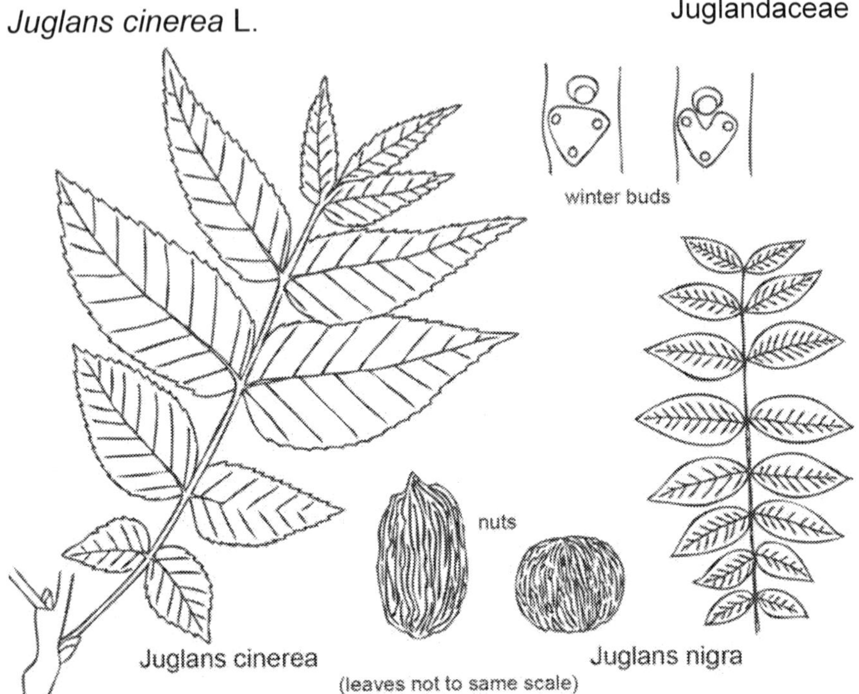

Flowers & Fruits: flowers monoecious catkins, both sexes unbranched; fruit, 4.0-8.0 cm, an elongate corrugated nut with thick, sticky, corky rind which disintegrates when ripe
Leaves: leaves feather compound-alternate, 30-60 cm long; leaflets 11-17, toothed, densely hairy beneath, crushed leaves spicy scented, end leaflets usually present; lateral leaflets progressively smaller toward base
Buds & Twigs: buds white-wooly; upper edge of the leaf scar not notched, with a hairy fringe; twigs with dark-brown chambered pith
Bark: dark; wide ridges, smooth-topped, separated by narrow fissures
Tree: to 20 m tall; diameter to 50 cm
Habitat: rich moist soil, river terraces and valleys, also dry slopes
Similar Species: the other trees with alternate-compound leaves almost always have fewer leaflets, a very different nut and their pith is not chambered; Black Walnut, *J. nigra,* can be difficult to distinguish as characters can overlap: it usually lacks an end leaflet, has light-brown chambered pith, has leaf scars that are notched, and (best character) has a round nut

Virginia Creeper

Parthenocissus quinquefolia Michx.

Status: FACU
Vitaceae

Flowers & Fruits: 5-parted, pale-green, small, in long-stemmed clusters; fruit in grape-like clusters, blue to black berries when ripe
Leaves: alternate-compound; 5 palmate leaflets, leaflets 6-12 cm, pointed, sharply toothed, shiny above, dark green in summer to red in fall
Buds & Twigs: light brown, slender, white pith, red lenticels; tendrils opposite the leaves, tendrils end in small adhesive disks
Bark: tight
Woody Vine: climbing or sprawling, up to 15 m tall
Habitat: swamps, edge of ponds and lakes, roadsides, streambanks.
Similar Species: Grape Woodbine, *P. vitaceae*, is similar but the tendrils do not end in adhesive disks (they curl around objects in the same way grape tendrils do) and the flower/fruit clusters are split into two equal branches

Swamp Rose

Rosa palustris (L.) Waldenb.

Status: OBL
Rosaceae

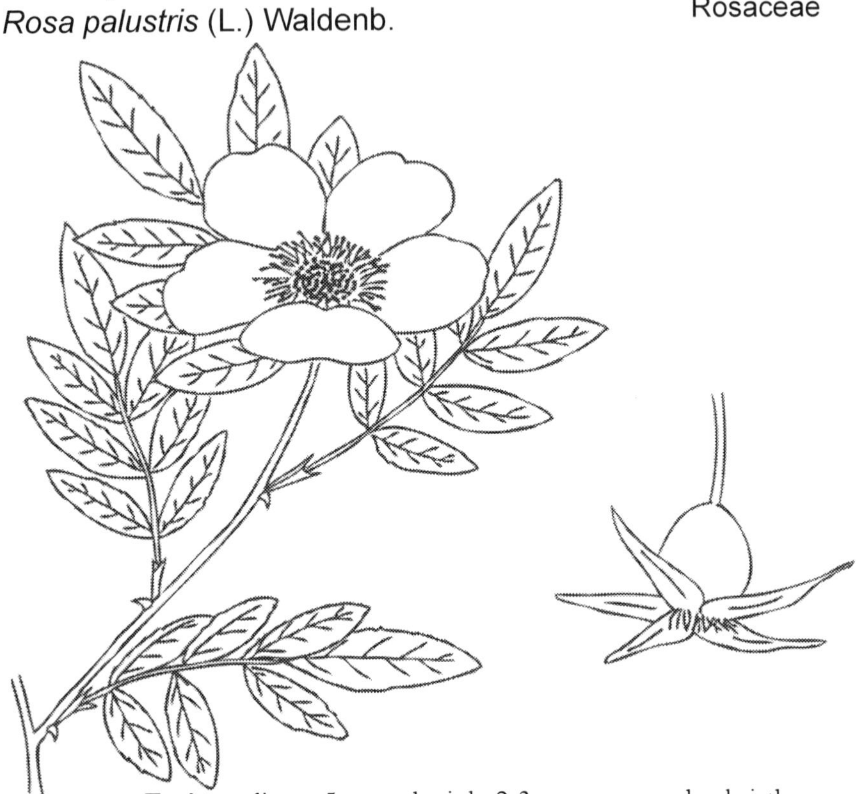

Flowers & Fruits: solitary; 5-parted, pink, 2-3 cm across, calyx bristly, fragrant; fruits bright red hips, 7-12 mm in diameter
Leaves: alternate-compound; 7 leaflets, elliptic, up to 4 cm long, pointed at tips, tapering to base, finely toothed
Buds & Twigs: stout (3-6 mm long, ~ half as wide as long) prickles at base of each stipule; no prickles between nodes of leaves
Shrub: much branched; up 2.5 m tall
Habitat: marshes, swamps
Similar Species: in the Adirondacks only Swamp Rose and Virginia Rose, *R. virginiana*, have prickles at the bases of the stipules but not between the leaf nodes; however Virginia Rose has coarsely toothed leaves
Other native species in the Adirondacks are Smooth Rose, *R. blanda*; Prickly Rose, *R. acicularis*; Wild Rose, *R. arkansana*; Carolina Rose, *C. carolina*; exotic species are Multiflora Rose, *R. multiflora* and Wrinkled Rose, *R. rugosa*

Common Blackberry

Rubus allegheniensis Porter

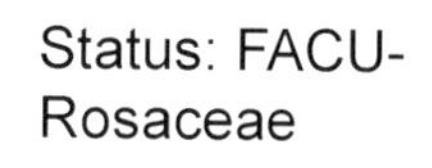
Status: FACU-
Rosaceae

Flowers & Fruits: flowers with 5 white petals, 5 sepals; fruit a cluster of drupelets, black, sweet when ripe; stalk of flower/fruit with glandular hairs
Leaves: alternate, fan-compound; 3-7 leaflets, undersides green
Buds & Twigs: leaf bases produce biennial canes which flower and fruit the second year; ridged stems, without whitish powder covering, with short, sometimes hooked thorns (usually enlarged at base), thorns also present on leaf stalks and even on veins on undersides of leaves
Shrub: erect arching canes lie flat on the ground; to 3 m long
Habitat: wide range of disturbed habitats
Similar Species: Pennsylvania Blackberry, *R. pensilvanicus*, is very similar to Common Blackberry but the stalk of the flower/fruit is not covered with glandular hairs; Black Raspberry, *R. occidentalis*, is also similar but the undersides of the leaflets are gray-white hairy
Red Raspberries, also have a stem with upright arching canes but they are covered with bristles rather than thorns; Dewberries have sprawing canes which trail on the ground

Bristly Dewberry
Rubus hispidus L.

Status: FACW
Rosaceae

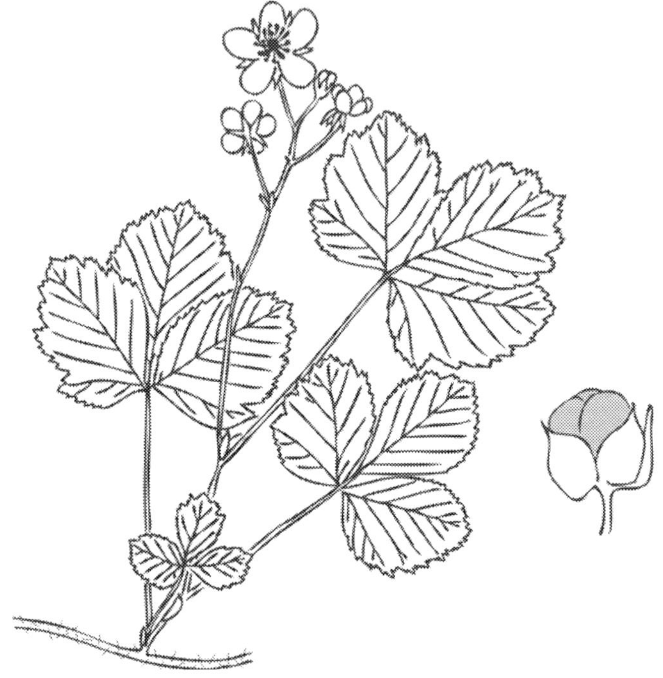

Flowers & Fruits: flowers with 5 white petals, 5 sepals; fruit a cluster of drupelets, sour when ripe
Leaves: alternate, fan-compound; 3-5 leaflets, undersides green
Buds & Twigs: leaf bases produce biennial canes which flower and fruit the second year; stems round, lacking whitish powder covering, with flexible, slender, bristles (2-4 mm long) which are usually not enlarged at base
Shrub: trailing canes lie flat on the ground; to 30 cm tall
Habitat: wide range, including swamps and peat bogs
Similar Species: Northern Dewberry, *R. flagellaris*, has stout thorns with enlarged bases on the stem; see Common Blackberry for other *Rubus spp* in the Adirondacks

Red Raspberry

Rubus idaeus L.

Status: FAC-

Rosaceae

Flowers & Fruits: flowers with 5 white petals, 5 sepals; fruit a cluster of drupelets; red, sweet when ripe

Leaves: alternate, fan-compound; 3-7 leaflets, undersides white-gray hairy (some keys consider tthese opposite-simple leaves)

Buds & Twigs: bases produce biennial canes which flower and fruit the second year; stems round, covered with whitish powder and flexible, slender, bristles (2-4 mm long) which are usually not enlarged at base

Shrub: erect arching canes; to 2 m tall

Habitat: wide range, wet and moist woods, thickets, roadsides

Similar Species: the only Red Raspberry in the Adirondacks; see Common Blackberry for other *Rubus spp* in the area

Mountain Ash

Sorbus americana Marsh.

Status: FACU
Rosaceae

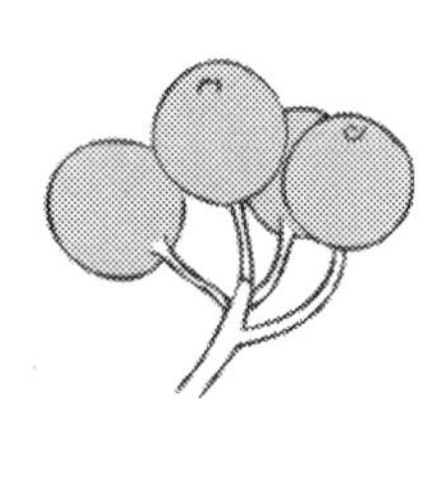

Flowers & Fruits: small, white, in clusters, 7.5-13 cm across; fruits bright red to orange-red berries
Leaves: alternate, compound leaves, 15-30 cm long; 11-17 lance-shaped leaflets, 5-7.5 cm long, length more than 3 x the width; dark green above, pale green below
Buds & Twigs: twigs gray to reddish brown, pith solid, narrow leaf scars; buds dark purplish red, 8-12 mm long, sticky
Bark: smooth with lenticels when young, grayish brown, wtih cracks, splits and scaly patches with age.
Shrub: to 10 m tall
Habitat: swamps, roadsides
Similar Species: Northern Mountain Ash, *S. decora*, has wider leaflets - length less than 3 x the width

Poison Ivy

Toxicodendron radicans (L.) Kuntze

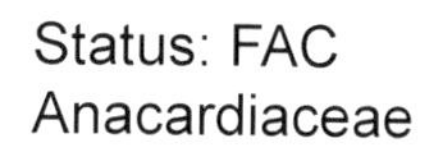

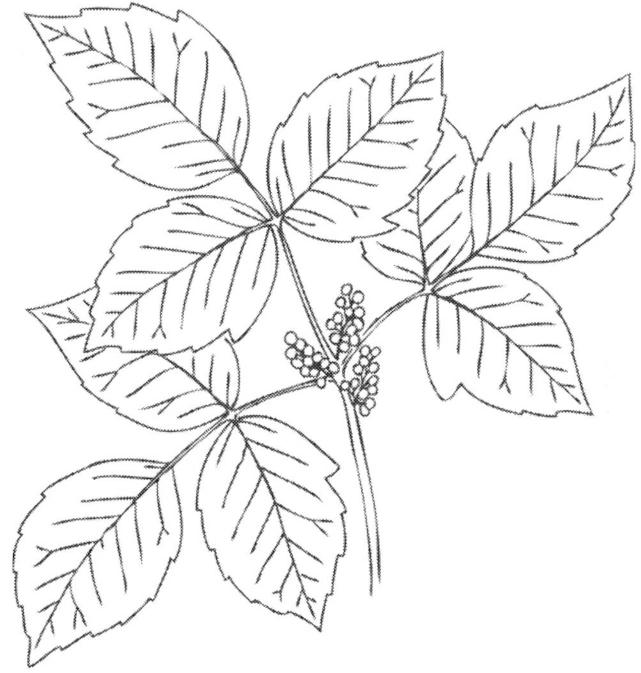

Flowers & Fruits: 5-parted flowers in racemes of 25 or more growing out of the axils of the leaves; fruits greenish-white berries, 3-5 mm

Leaves: alternate-compound; 3 leaflets (5-15 cm), often irregularly shaped with lobes and/or teeth; terminal leaflet long petioled, lateral leaflets sessile or short petioled; young leaflets often glossy and reddish tinted

Buds & Twigs: buds hairy; twigs brownish; leaf scars crescent-shaped; aerial rootlets attach vine to substrate

Habit: usually occurs as a climbing vine, sometimes overcomes original substrate and becomes free-standing; TOXIC - poisonous to touch, causes rash

Habitat: woods, thickets, fields; especially in disturbed areas

Similar Species: Western Poison Ivy, *T. rydbergii*, does not climb and has no aerial rootlets; Virginia Creeper, *Parthenocissus quinquefolia*, and Grape Woodbine, *P. vitaceae* have 5 leaflets

Poison Sumac

Toxicodendron vernix (L.) Kuntze

Status: OBL
Anacardiaceae

Flowers & Fruits: small, 3 mm long, yellow to green, in loose clusters; small, round, white to pale green berries in clusters
Leaves: alternate, compound, 20-33 cm long; 7-13 ovate leaflets, 5-10 cm long, shiny green above, pale green beneath
Buds & Twigs: orange brown, flat-sided, with dark brown lenticels, large shield-shaped leaf scars; dark colored sap, very toxic
Bark: smooth, grayish brown with dark brown lenticles
Shrub: to 5 m tall; TOXIC - poisonous to touch, causes rash
Habitat: swamps, marshes, edge of lakes and ponds
Similar Species: twigs of Smooth Sumac, *Rhus copallinum*, are similar but the twigs are round and the fruits are red; twigs of Staghorn Sumac, *R. typhina*, are velvety, covered with reddish brown hairs, and the fruits are red; Fragrant Sumac, *R. aromatica*, has only 3 leaflets

Ninebark

Physocarpa opulifolius (L.) Maxim

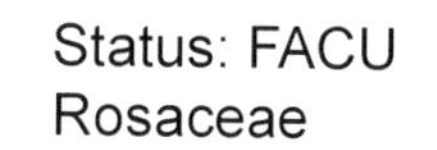

Flowers & Fruits: 5-parted, white to pale pink to pink, 7-10 mm; in corymbs, numerous; 25-50 mm wide; fruits red in clusters of 3-5, 5-12 mm wide

Leaves: alternate, simple, ovate-orbicular, 3-5 palmate lobes, toothed margin; 25-75 mm long, green to burgundy (in fall)

Buds & Twigs: yellow to organge to brown, slender

Bark: darker older bark splits and exfoliates in long strips

Shrub: up to 3 m tall

Habitat: streambanks, moist soil along shores

Sycamore

Platanus occidentalis L.

Status: FACW-
Platanaceae

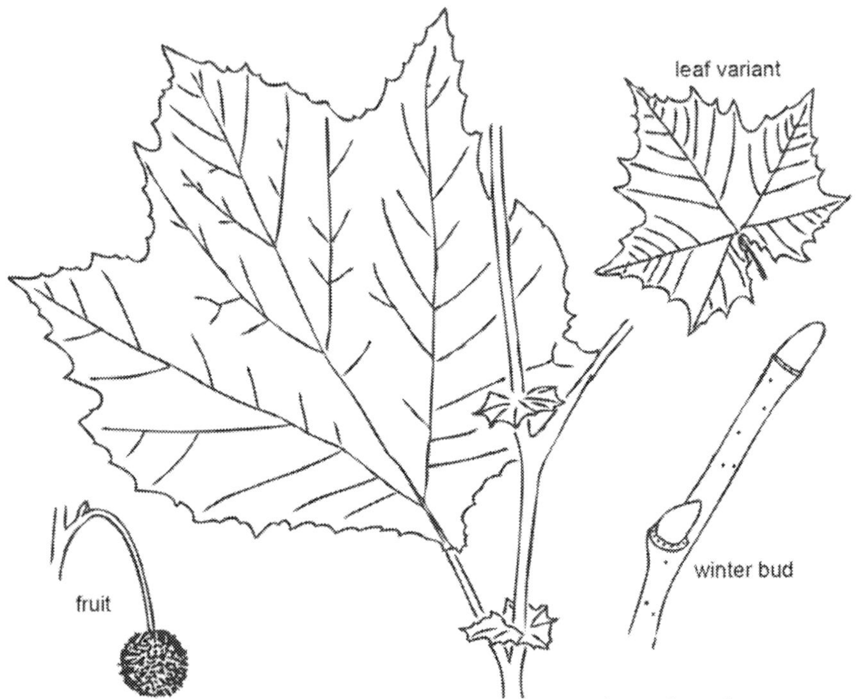

Flowers & Fruits: plants usually monoecious, sometimes found on different branches of the same tree; flowers in tiny spherical heads; fruits in a spherical head (2.5-3.0 cm) of seeds, each 7-8 mm, with many long hairs at the base
Leaves: alternate-simple, 3-5 lobes with fan-shaped veins, toothed, 15-25 cm; leaf stalk hollow, leaf like stipule at base of stalk
Buds & Twigs: 1 bud scale; leaf scars with ring extending around twig
Bark: pale gray to white-green bark exfoliates in patches producing a mottled bark
Tree: to 50 m tall; diameter to 3 m
Habitat: moist, wet, alluvial soil
Similar Species: none - the leaves, bark, and fruits are all distinctive

American Black Current

Ribes americanum (P.) Mill.

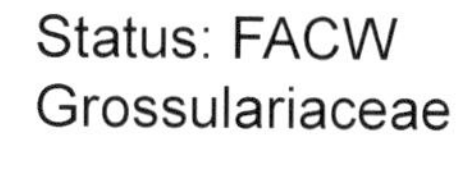

Status: FACW
Grossulariaceae

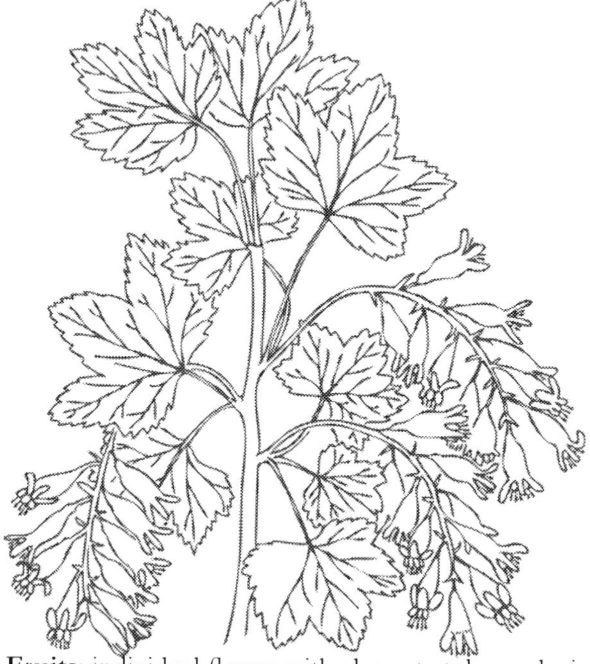

Flowers & Fruits: individual flower with elongate tube enclosing the ovary; 5-parted; sepals form petal-like lobes at the end of the tube; true petals small; tube and sepals yellow-white; flowers in a raceme; fruit a black berry covered with resin-dots
Leaves: alternate-simple, 1.5-10.0 cm; 3-5 deeply cleft lobes, fan-veined; undersides of leaves with numerous tiny yellow resin-dots (use lens)
Buds & Twigs: leaf scar with ridge descending from the center of the scar; broken twigs odorless
Shrubs: erect shrub; to 1.5 m
Habitat: floodplains, woods openings
Similar Species: in addition to American Black Current only Skunk Current, *R. glandulosum*, and Swamp Current, *R. triste*, have stems without prickles and/or thorns: both of these have red fruits and neither has numerous resin-dots on the undersides of the leaves; four other *Ribes* in the Adirondacks have a thorny and/or prickly stem: Prickly Gooseberry, *R. cynosbati*; Smooth Gooseberry, *R. hirtellum*; Bristly Black Current, *R. lacustre*, and Appalachian Gooseberry, *R. rotundifolium*

Purple Flowered Raspberry

Status: unknown
Rosaceae

Rubus odoratus L.

Flowers & Fruits: flowers rose-purple petals, yellow stamens & pistils, 2-5 cm diameter; in a loose, branched cluster; fruit 1 cm, tasteless, druplets falling individually
Leaves: alternate simple, 10-20 cm wide; 3-5 triangular lobes, fan-veined, sharply toothed; usually hairy beneath; leaves break off at base of leaf leaving leaf stalks on cane
Buds & Twigs: twigs hairy
Bark: older bark papery, shreddy
Shrub; flexible canes, to 2 m tall
Habitat: shady openings, edges of fields
Similar Species: the only species of *Rubus* that does not have compound leaves; see also American Black Current, *Ribes americanum*

Basswood

Tilia americana L.

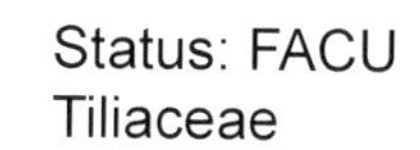

Status: FACU
Tiliaceae

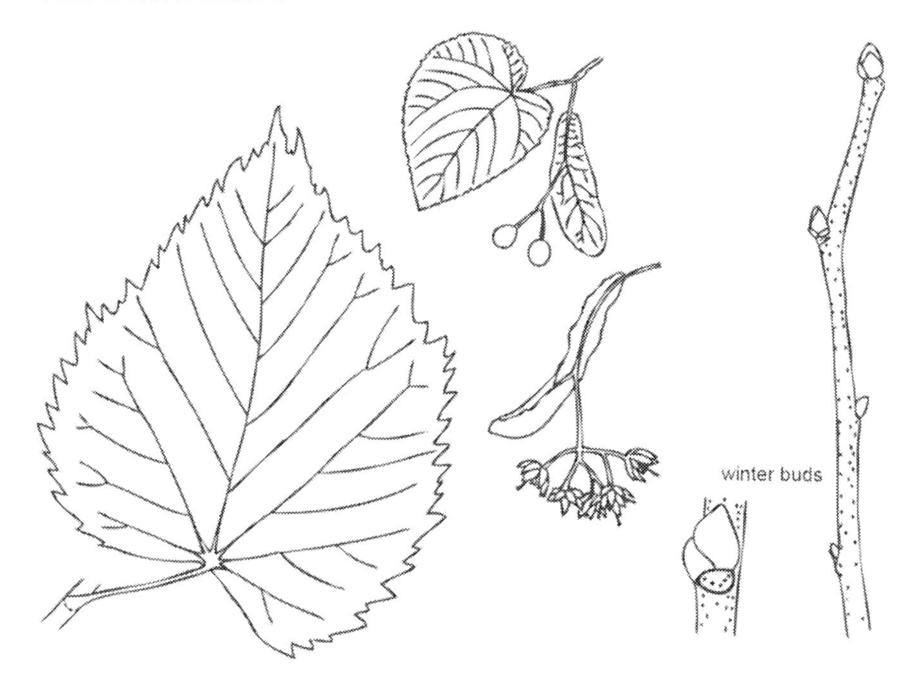

Flowers & Fruits: flowers 5-parted, yellow-white petals 7-12 mm, fragrant; in clusters in the axils of leaves on a long stalk that is fused, to about the midpoint, to an elongate leafy bract; blooms mid-summer; fruit nut-like
Leaves: alternate-simple; 7-15 cm; heart-shaped, bases very uneven, an elongate tip; fan-veined (3-5 large veins meeting near the base of the leaf), fine-toothed, never lobed
Buds & Twigs: buds bright red; twigs and smaller branches gray, smooth
Bark: gray, deep grooves separating long narrow ridges
Tree: to 40 m tall; often grows in clumps from sprouts after logging or fires
Habitat: moist woods
Similar Species: the shape of the leaves is fairly distinctive; the flowers and fruit more so

Riverbank Grape

Vitis riparia Michx.

Status: FACW
Vitaceae

Flowers & Fruit: panicle of 4-parted flowers, panicles 5-10 cm; juicey berry blue-black with heavy waxy powder, 6-12 mm, bitter
Leaves: alternate-simple, 10-20 cm; leaf shape variable, somewhat 3-lobed, sharply toothed (teeth longer than wide with one or both sides concave), longer than broad; shiny green above, duller green below; pubescent beneath when young, permanently so in vein-axils and along veins
Buds & Twigs: twigs hairless; pith partitioned (0.5-2.0 mm) opposite leaf base; branched tendrils and flower/fruit clustes arise opposite leaves, absent every third leaf
Bark: dark brown, shreds into long strips
Woody Vine: using tendrils climbs high in trees
Habitat: moist woods, roadsides, thickets
Similar Species: Winter Grape, *V. vulpina*, is the most similar but has broader duller teeth (teeth wider than long and both sides convex); Summer Grape, *V. aestivalis*, Fox Grape, *V. labrusca*, and New England Grape, *V. novae-angliae*, all have reddish-wooly hairs on the undersides of the leaves

Bog rosemary

Andromeda poliphylla L.

Status: OBL
Ericaceae

Flowers & Fruits: small terminal clusters of nodding white to pinkish-white flowers, 5 corolla lobes fused into a nearly cylindrical bell-like shape, 5-6 mm; fruit a dry capsule
Leaves: alternate-simple; 2.5-4.0 cm; leaves narrow-elongate, not toothed, leathery, evergreen; dark green above, undersides whitened by tiny erect hairs
Twigs & Buds: hairless
Shrub: height 30-60 cm
Habitat: acid bogs
Similar Species: two other bog shrubs have alternate simple leaves: in Leatherleaf, *Chamaedaphne calyculata,* the distal leaves become much smaller than the proximal ones, the leaf margins are not inrolled, and the undersides of the leaves have yellow waxy deposits; in Labrador Tea, *Ledum groenlandicum*, has wider leaves with long rusty to whitish hair on the undersides, and hairy twigs (the flowers, if present, are also very different)
aka: *A. glaucophylla*

Leatherleaf

Chamaedaphne calyculata (L.) Moench

Status: OBL
Ericaceae

Flowers & Fruits: flowers 5-parted, corolla lobes fused into nearly cylindrical bell shape, 6-7 mm, whitish; in axils of current year's leaves, forming a 1-sided raceme

Leaves: alternate-simple, 1.5-5.0 cm ; leaves, elliptical, toothless, leathery, evergreen (stay on shrub 2 years), the current year's continue to be formed throughout summer so that youngest and smallest are at the very tip of the twig; undersides of leaves with scaley yellow wax deposits (use lens)

Twigs & Buds: surface with scaley yellow wax deposits (use lens)

Shrub: height to 1 m

Habitat: acid bogs and margins of small nutrient-poor ponds

Similar Species: of the leathery leafed plants in this habitat, Sheep Laurel, *Kalmia angustifolia* , and Bog Laurel, *K. polifolia*, both have opposite or whorled leaves; of those with alternate leaves Labrador Tea, *Ledum groenlandicum*, has rolled leaf edges and is rusty- or white- wooly underneath and Bog Rosemary, *Andromeda glaucophylla*, has narrow, rolled-edge leaves whitened underneath by tiny hairs

Leatherwood

Dirca palustris L.

Status: FAC

Thymelaeaceae

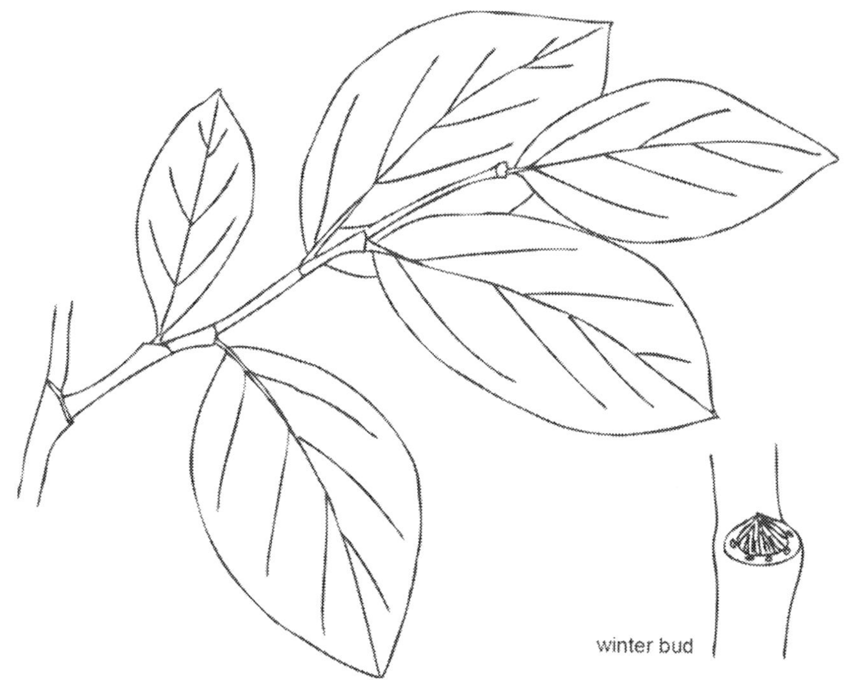

Flowers & Fruits: base of sepals fused to form a tube enclosing flower; petals minute, pale yellow-green; fruit turns from green to red, 12-15mm

Leaves: alternate-simple, 5-8 cm; ovate, entire; short leaf stalks (5-8 mm)

Buds & Twigs: short leaf stalks entirely cover velvety-brown buds; leaf scars raised well above surface of twig, leaf scars encircle next year's bud making the twig appear jointed, 5 bundle scars; bark of twigs very tough and fibrous, cannot be broken by hand

Shrub: to 2 m

Habitat: rich, moist woods

Similar Species: the raised leaf scars with 5 bundles and the tough bark distinguish the species from others

Smooth Winterberry Holly

Status: OBL

Ilex laevigata (Pursh) Gray

Aquifoliaceae

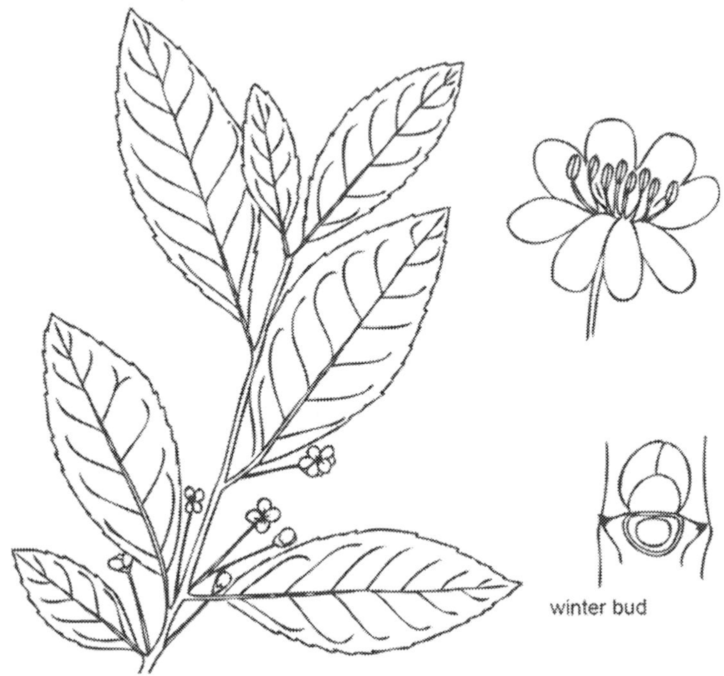

Flowers & Fruits: flowers unisexual, 4 or 8 parted; calyx small, individual sepals with a smooth edge (this characteristic remains visible in the fruit); flowers (and fruits) in small clusters, each on a moderately long stalk; fruit a red berry containing a smooth nutlet

Leaves: alternate-simple, 4-8 cm; variable shape from narrow to egg-shaped, widest near the center; fine teeth, shiny upper surface, smooth beneath

Buds & Twigs: buds blunt, scales sharply pointed; leaf scars flanked by tiny black thorn-like stipules (use lens); twigs light brown

Bark: mottled gray with few warty lenticils

Shrub: to 3 m tall

Habitat: wet thickets, swamps

Similar Species: the small dark thorny stipules distinguish the genus *Ilex* from others; Smooth Winterberry Holly, *I. verticillata,* has sepal edges that are fringed (the best character), fruits and flowers on shorter stalks, and bark with more warty lenticils

Winterberry Holly

Ilex verticillata (L.) Gray

Status: FACW+
Aquifoliaceae

Flowers & Fruits: flowers unisexual, 4 or 8 parted; calyx small, individual sepals with a fringed edge (this characteristic remains visible in the fruit); flowers (and fruits) in small clusters, each on a short stalk; fruit a red berry containing a smooth nutlet

Leaves: alternate-simple, 4-8 cm; variable shape from egg-shaped to almost circular (1.0 to 2.5 times as long as wide), widest near the upper end; distinct coarse teeth, dull upper surface

Buds & Twigs: buds blunt, scales broadly pointed; leaf scars flanked by tiny black thorn-like stipules (use lens); twigs light brown

Bark: mottled gray with warty lenticils

Shrub: to 4 m tall

Habitat: wet thickets, swamps

Similar Species: the small dark thorny stipules distinguish the genus *Ilex* from others; Smooth Winterberry Holly *I. laevigata* has sepal edges that are not fringed (the best character), fruits and flowers on longer stalks, and bark with few warty lenticils

Labrador Tea

Ledum groenlandicum Oeder.

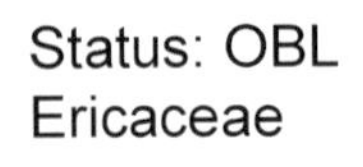

Flowers & Fruits: small (~1 cm), 5-parted, white, in terminal clusters
Leaves: alternate-simple; narrow, 20-50 mm long, inrolled edges, toothless, fragrant when crushed; dark green above, densely white- or rusty-hairy beneath
Twigs: densely hairy
Shrub: to 100 cm
Habitat: cold bogs shores, peat soils, sometimes on rocky alpine slopes
Similar Species: leaves similar to Bog Laurel, *Kalmia poliphylla,* but that species has opposite leaves; Bog Rosemary, *Andromeda glaucophylla,* has alternate leaves with inrolled edges, but the undersides are whitened by short erect hairs; Leatherleaf, *Chamaedaphne calyculata,* has yellowish waxy deposits on the undersides of the leaves and the leaves become progressively smaller toward the end of the current year's growth

Mountain Holly

Status: OBL

Nemopanthus mucronatus (L.) Loes.

Aquifoliaceae

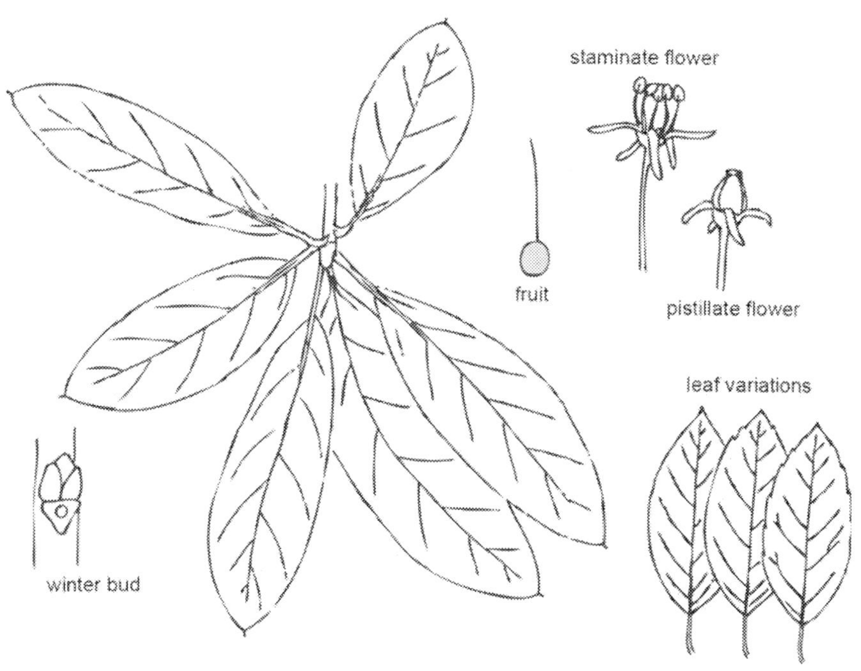

Flowers & Fruits: flowers unisexual (usually on separate plants), occasionally bisexual, 4-5 parted; calyx small, petals yellow; flowers (and fruits) in leaf axils, each on a long stalk; fruit a velvety red berry (~ 6 mm);
Leaves: alternate-simple, 3-7 cm; oblong (often sides almost parallel) with a tiny bristle tip; fine teeth or (usually) none; leaf stalks often purplish; single vascular bundle in each leaf scar
Buds & Twigs: buds small, ovoid; twigs often purplish
Bark: smooth gray
Shrub: to 5 m tall
Habitat: wet thickets, swamps, edges of small ponds and streams
Similar Species: the purplish twigs and leaf stalks, bristle-tips on the oblong leaves, and the velvety red fruit distinguish this species from others

Rhodora

Rhododendron canadense (L.) Torr.

Status: FACW
Ericaceae

Flowers & Fruits: 5-parted, pink to pinkish purple, in terminal clusters of 2-6; 2-3 cm across, 3 upper lobes erect, 2 lower lobes are oblong and spreading; flowers in spring before leaves open
Leaves: alternate, simple, grayish green, 2-6 cm long, blunt-tipped, oval, hairy underneath
Buds & Twigs: buds light gray, round, hairy, 1 cm in diameter; flower buds have only a few scales; twigs hairless
Bark: dark gray to black, smooth
Shrub: up to 2 m tall
Habitat: bogs, fens
Similar Species: other *Rhododendron* in the Adirondacks don't have blunt leaves, hairy buds, hairless twigs, and few-scaled flower buds

Swamp Azalea

Rhododendron viscosum (L.) Torr.

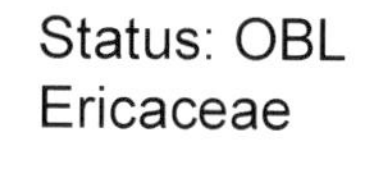

Status: OBL
Ericaceae

Flowers & Fruits: radial symmetry; 5-parted, white to pale pink with sticky reddish hairs, 25-40 mm long; sweet scented; appear in summer after leaves opened

Leaves: alternate, simple, clustered at the end of branches, 3-6 cm long, ends pointed to slightly rounded, glossy green above, white hairless beneath

Buds & Twigs: buds hairy; twigs bristly-hairy; flower buds with 8-12 scales

Shrub: up to 2 m tall

Habitat: swamps

Similar Species: Great Laurel, *R. maximum*, and Lapland Azalea, *R. lapponicum*, have thick leathery evergreen leaves; Early Azalea, *R. prinophyllum* has bright pink flowers opening early in the spring, with or before the leaves

Lowbush Blueberry (Late Lowbush)

Vaccinium angustifolium Ait.

Status: FACU
Ericaceae

Flowers & Fruits: flowers in axillary or terminal racemes; corolla tubular with shallow lobes, 4-6 mm long, white to pinkish white; fruit a berry, blue, covered with a white waxy bloom, 5-10 mm diameter
Leaves: alternate-simple; elliptical; deciduous; sharply toothed; 1.5-3.0 cm, 2-3 times as long as wide; sometimes hairy or whitened beneath
Buds & Twigs: twigs hairy or hairless, covered with tiny warts (use lens)
Shrub: low, much-branched, very colonial, to 30 cm high
Habitat: moist or dry woods, well-drained sandy or rocky soil
Similar Species: most similar to Early Lowbush Blueberry (aka Hillside Blueberry), *V. pallidum,* which has longer (3-5 cm), relatively wider, leaves, and flowers/fruits earlier in the season; Velvetleaf Blueberry, *V. myrtilloides,* has leaves the undersides covered with soft hairs; other species also found in the Adirondacks are: Dwarf Bilberry, *V. cespitosum*; Northen Blueberry, *V. boreale*; Deerberry, V. *stamineum*; and Bog Blueberry, *V. uliginosum*

Highbush Blueberry

Vaccinium corymbosum L.

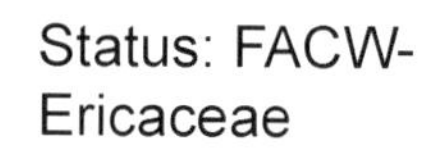

Flowers & Fruits: flowers in axillary or terminal racemes; corolla tubular with shallow lobes, 5-10 mm long, white to pinkish white; fruit a berry, blue to black, 5-12 mm diameter
Leaves: alternate-simple; egg-shaped to elliptical;deciduous; toothed or toothless; underside hairless or sparsely hairy, sometimes with fine waxy covering; 3.0-8.0 x 1.5-4.0 cm, twice as long as wide
Buds & Twigs: twigs hairy or hairless, covered with tiny warts (use lens)
Shrub: to 3 m high
Habitat: open swamps and bogs, upland fields, old woods, mountain tops
Similar Species: only highbush blueberry in the Adirondacks; Early Lowbush Blueberry, *V. pallidum,* and small Highbush Blueberry are similar and their range of variability overlaps; distinguishable mainly by size in older flowering/fruiting specimens, but smaller specimens can cause difficulty

Speckled Alder
Alnus incana (L.) Moench

Status: Unknown
Betulaceae

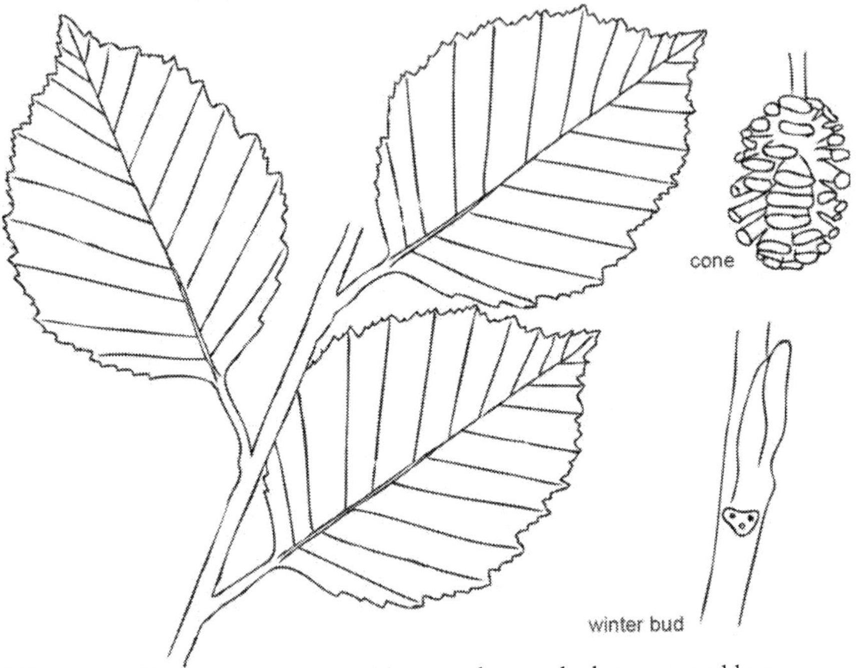

cone

winter bud

Flowers & Fruits: staminate catkins are clustered, elongate, and hang downward; pistillate catkins are clustered, shorter, short-stalked, woody, hang downward, and usually persist on shrub through the following winter
Leaves: alternate-simple, 5-12 cm, usually droop downward; egg-shaped, usually broadest below the middle, base rounded to heart-shaped; sharply double or single-toothed, almost with lobes; upper surface wrinkled, dull-green, veins deeply impressed; under surface hoary, veins conspicuously projecting below, veinlets forming ladder-like pattern
Buds & Twigs: buds on short stalks; pith triangular
Bark: dark, speckled with numerous transverse white lenticils
Shrub: to 4 m tall; diameter to 10 cm
Habitat: streambanks, swamps, wet soils
Similar Species: Smooth Alder, *A. serrulata,* and Green Alder, *A. viridis*, haves bark with only a few of the white lenticils and leaves that tend to be broadest above the middle; in both the pistillate catkins, fruits, and leaves tend to be erect

Smooth Alder

Alnus serrulata (Ait.) Willd.

Status: OBL
Betulaceae

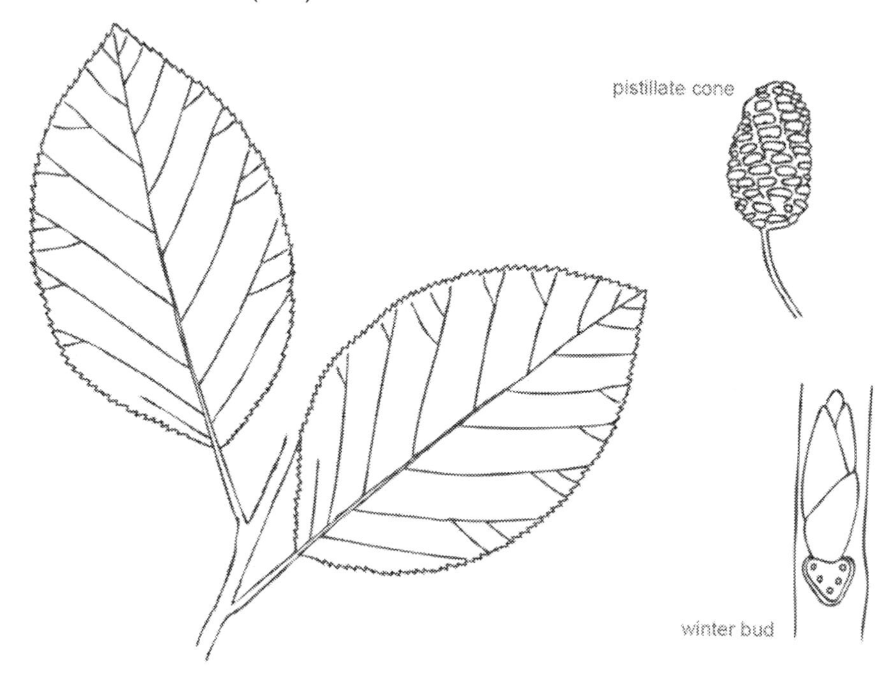

Flowers & Fruits: staminate catkins clustered, elongate, pendent; pistillate catkins also clustered, short-stalked, erect, and woody - usually persist on shrub through the following winter
Leaves: alternate-simple, 5-12 cm, erect; egg-shaped, usually broadest at or above the middle, base wedge-shaped; sharply double or single-toothed; shiny green on upper surface; yellow-green beneath
Buds & Twigs: buds on short stalks; pith triangular
Bark: dark, speckled with a few short transverse white lenticils or without
Shrub: to 4 m tall; diameter to 10 cm
Habitat: streambanks
Similar Species: in Green Alder, *A. viridis*, the leaf margin is more likely to be single toothed than double-toothed and the base of the leaf is more rounded; in Speckled Alder, *A. incana,* has bark with numerous white lenticils and leaves that tend to be broadest below the middle; both the pistillate catkins, fruits, and the leaves tend to hang downward rather than be erect

Downy Juneberry

Amelanchier arborea (Michx. f.) Fernald

Status: Unknown
Rosaceae

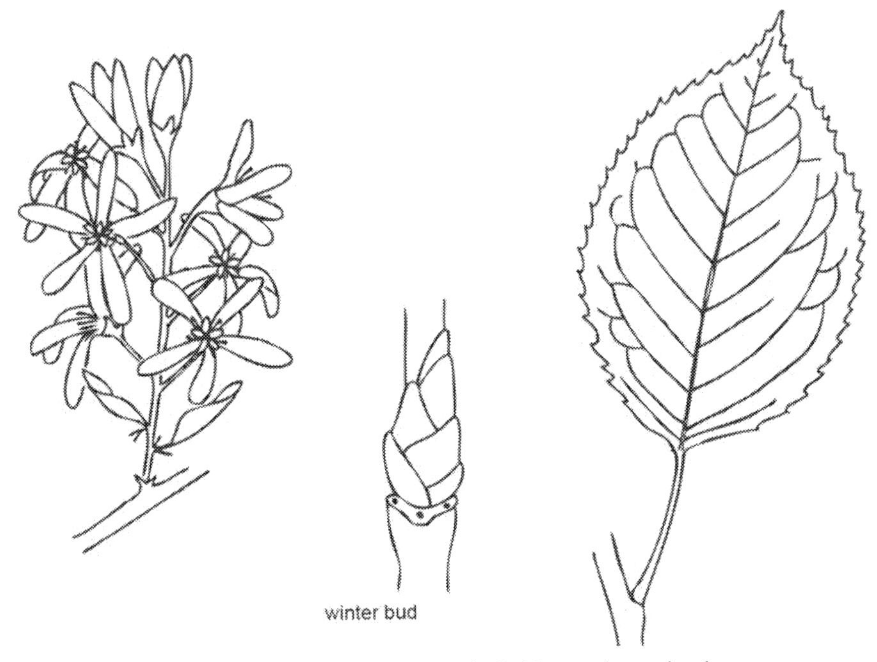

Flowers & Fruits: 5-parted, white; petals 8-12 mm long; in dense erect clusters, flowers very early spring; fruits purple-black, dry, flavorless; lowermost fruitstalk ~12 mm long
Leaves: alternate-simple; oval, 5-9 cm long; tapered to a sharp tip; about 25 teeth per side with about 12 veins per side, veins stop short of teeth; at flowering time leaves folded, less than half grown, densely hairy when new but lost late during season
Buds & Twigs: buds narrow, tapering to a point; leaf scars with 3 large bundles; slender with ridges extending downward from each side of leaf scar; pith 5-pointed in cross-section
Tree: shrubs or small trees to 12 m
Habitat: dry sites in wooded areas and along edges
Similar Species: Smooth Juneberry, *A. laevis,* is very similar but the leaves are at least half grown at flowering and are coppery red and hairless, the petals are longer (10-17 mm) and in drooping clusters; when the fruits are present they are sweet and juicy and the stalk of the lowermost fruit is 10-17 mm; during summer they are difficult to distinguish

Yellow Birch

Status: FAC

Betula alleghaniensis Britt.

Betulaceae

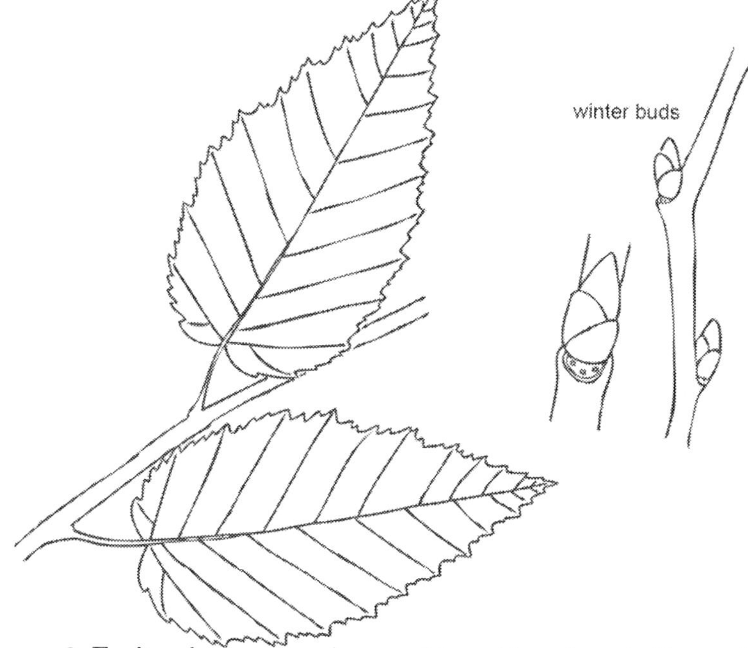

Flowers & Fruits: elongate staminate catkins with scales; pistillate catkins short, ovoid, scales 5-7 mm; fruit with 2 lateral wings
Leaves: alternate-simple, 6-10 cm; double toothed; ovate, rounded at base; hairless; main side veins mostly not branched, each side vein ending at the tip of a tooth
Buds & Twigs: somehat hairy; broken twigs with a moderately intense sweet wintergreen odor
Bark: bark of larger trunks dirty yellowish white, peels in narrow shaggy strips; bark of younger branches shiny yellow-brown with narrow cross stripes
Tree: to 30 m; diameter to 1 m
Habitat: moist, especially secondary, forests
Similar Species: Yellow Birch is the only birch with yellowish bark; in both Paper Birch, *B. papyrifera*, and Gray Birch, *B. populifolia*, the bark is white; in Sweet Birch, *B. lenta*, the bark is dark reddish-brown, but small saplings of Yellow Birch can be difficult to distinguish - use the lack of branching in the main side veins of the leaves and the hairy buds and twigs

Paper Birch

Betula papyrifera Marsh.

Status: FACU
Betulaceae

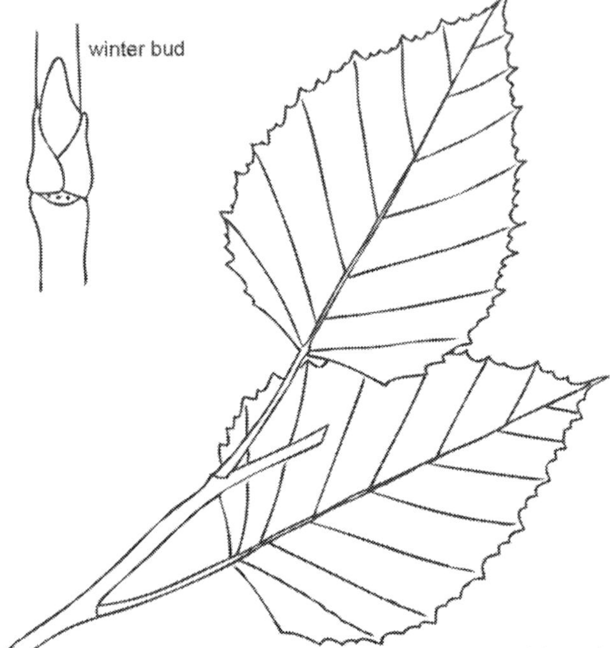

Flowers & Fruits: elongate (3-5 cm) staminate catkins with scales; pistillate catkins shorter, scales 2-3 mm; fruit with 2 lateral wings

Leaves: alternate-simple, 5-10 cm; double toothed; ovate, wedge-shaped or rounded at base; 9 veins per side or fewer, each ending at tip of tooth; almost hairless except in axils of veins

Buds & Twigs: somewhat hairy, 5-7 mm; broken twigs without sweet wintergreen odor

Bark: bark creamy white with medium sized horizontal to semilunate black marks around side branches; peels in large plates exposing salmon-pink inner side; narrow horizontal streaks in bark

Tree: to 20 m; diameter to 60 cm; often leaning

Habitat: dry or moist soil in secondary woods

Similar Species: Yellow Birch, *B. alleghaniensis*, does not have truly white bark and twigs have a wintergreen smell when broken; Gray Birch, *B. populifolia*, has white bark but it does not peel in large plates, its twigs are rough-warty, and the leaves are triangular in shape rather than ovate

Gray Birch

Status: FAC

Betula populifolia Marsh.

Betulaceae

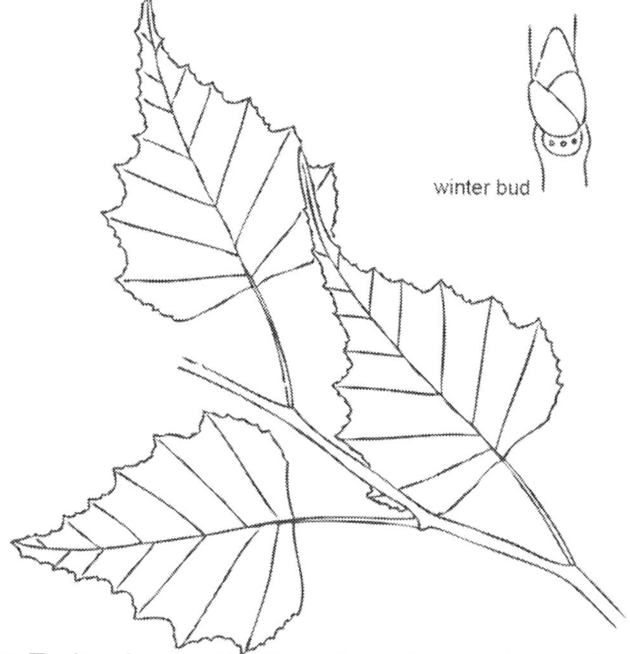
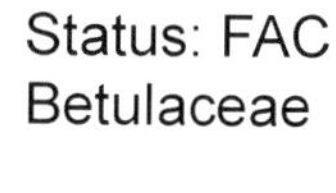

Flowers & Fruits: elongate (1.3-3.0 cm) staminate catkins with scales; pistillate catkins shorter; fruit with 2 lateral wings
Leaves: alternate-simple, 4-7 cm; double toothed except on base; long-pointed triangular, flattened at the base; pendulous; almost hairless except in axils of veins
Buds & Twigs: twigs rough-warty; broken twigs without sweet wintergreen odor
Bark: bark chalky white with large chevron shaped black marks around side branches; peels only in small strips; narrow horizontal streaks in bark
Tree: to 10 m; diameter to 15 cm; trunks commonly clustered
Habitat: poor soils in old fields and secondary woods
Similar Species: Yellow Birch, *B. alleghaniensis*, does not have truly white bark and its twigs have a wintergreen odor when broken; Paper Birch, *B. papyrifera*, has white bark that peels in large plates, has hairy twigs, and leaves are ovate in shape rather than triangular

American Hornbeam

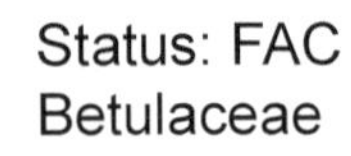

Carpinus caroliniana Walter

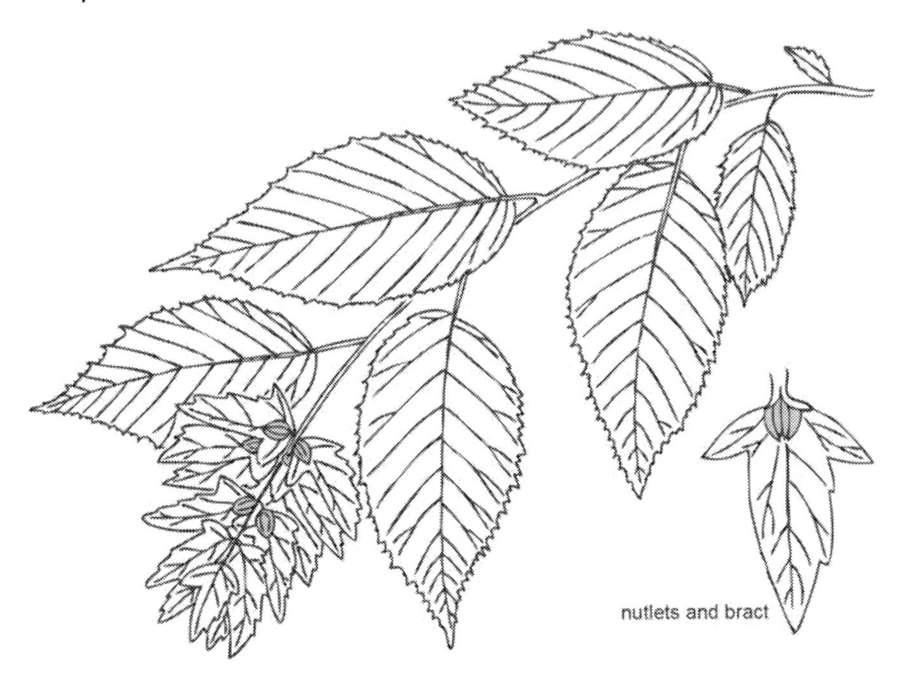

Flowers & Fruits: staminate and pistillate catkins on same tree; fruit pendent spike of tiny nuts attached to base of 3-pointed bracts
Leaves: alternate-simple, 5-10 cm; usually double-toothed, egg-shaped, with pointed tips, rounded at base; veins parallel, each ending at tip of tooth; become progressively larger along shoot
Buds & Twigs: end buds absent; twigs somewhat square in cross-section
Bark: smooth, gray; with muscule-like longitudinal ridges
Tree: to 10 m; diameter to 50 cm
Habitat: moist soils
Similar Species: bark and bracts of nuts are very distinctive; leaves similar to Hop Hornbeam, *Ostrya virginiana*
aka: Musclewood

Hazelnut

Corylus americana Walt.

Status: FACU-
Betulaceae

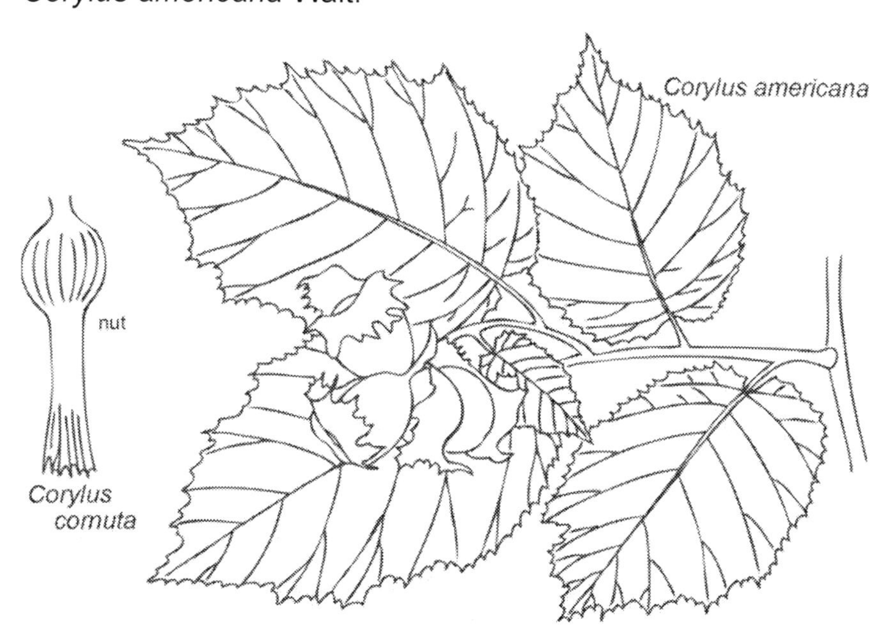

Flowers & Fruits: staminate flowers in catkin; fruit a nut (1.0-1.5 cm) enclosed by thin, flattened, ragged-edged husk
Leaves: alternate-simple, 5-12cm; double-toothed; somewhat heart-shaped, tip pointed, more or less hairy underneath
Buds & Twigs: buds, twigs, petioles of leaves bristly hairy; leaf-stalks and twigs with stalked glands; buds rounded, blunt, brown-gray
Bark: smooth, brown
Shrub: to 3 m tall
Habitat: dry or moist woods, thickets
Similar Species: Beaked Hazelnut, *C. cornutus*, has a husk with an elongate tube enclosing the nut and the twigs are without stalked glands

Sweetgale
Myrica gale L.

Status: OBL
Myricaceae

Flowers & Fruits: yellowish green, 10-12 mm, in catkins; flowering before leaves open
Leaves: alternate, simple, 25-75 mm long, dark green above, light green below, oblanceolate to obovate, tapering at the base, usually toothed at the tip; strongly aromatic, resin dots on undersides
Buds & Twigs: small pointed buds, dark brown; twigs slender, hairless, with resin dots
Bark: reddish brown with pale-brown lenticels
Shrub: up to 1.5 m tall
Habitat: bogs, swamps, edges of lakes and ponds
Similar Species: can be distinguished from Northern Bayberry, *M. pensylvanica*, which has aromatic wax-covered fruits and hairy twigs

Hop Hornbeam

Ostrya virginiana (Mill) K. Koch

Status: FACU-

Betulaceae

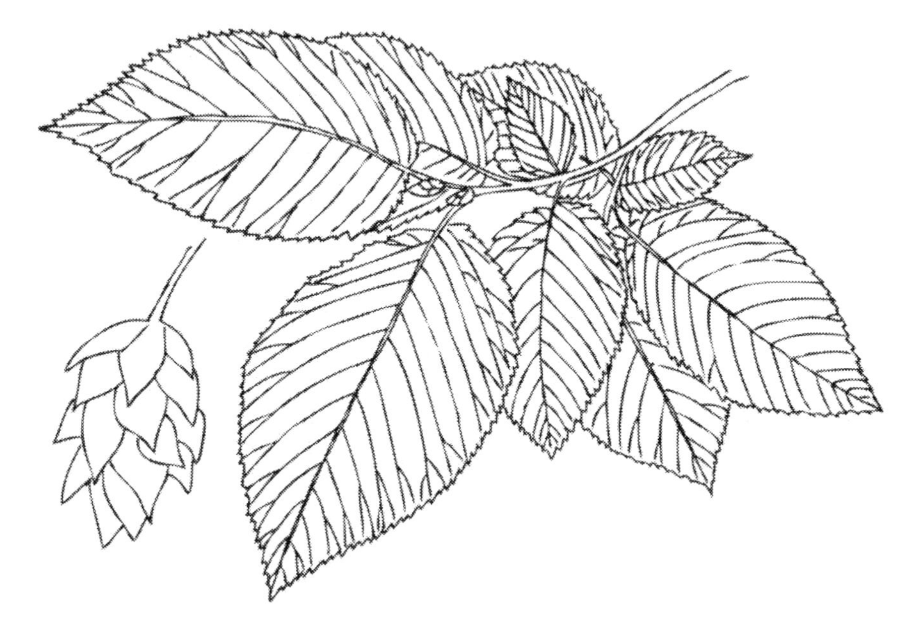

Flowers & Fruits: staminate catkins; fruit pendent spike of small bladder-enclosed nuts attached to base of bracts

Leaves: alternate-simple, 2.5-12.0 cm; usually double-toothed, egg-shaped, with pointed tips

Buds & Twigs: end buds absent, bud scales finely grooved (use lens)

Bark: brown, cracking into long loose shreddy, rectangular plates, rough

Tree: to 10 m; diameter to 50 cm

Habitat: moist to dry soils

Similar Species: leaves similar to American Hornbeam, *Carpinus caroliniana*, but the bark and bracts of nuts are very different

Black Chokeberry

Photinia melanocarpa (Michx.) K.R. Rob

Status: FAC
Rosaceae

Flowers & Fruits: flowers with 5 sepals and 5 white petals (4-6 mm), numerous red-anthered stamens; infloresence a rounded cluster; fruits black, 4-10 mm

Leaves: alternate-simple, 3-7 cm; elliptical with pointed tip, toothed; leaves with small black glands along the top of the midrib (use lens)

Buds & Twigs: buds reddish, scale tips slightly notched; twigs hairless

Shrub: to 3 m tall

Habitat: bogs, swamps, wetlands, also in drier thickets

Similar Species: the small black glands along the top of the midrib are unique to the Chokeberries; Red Chokeberry, *P. arbutifolia*, has twigs and leaf undersides that are hairy and red fruit; Purple Chokeberry, *P. floribunda*, a hybrid, is intermediately hairy and has purple fruits

Balsam Poplar

Populus balsimifera L.

Status: FACW
Salicaceae

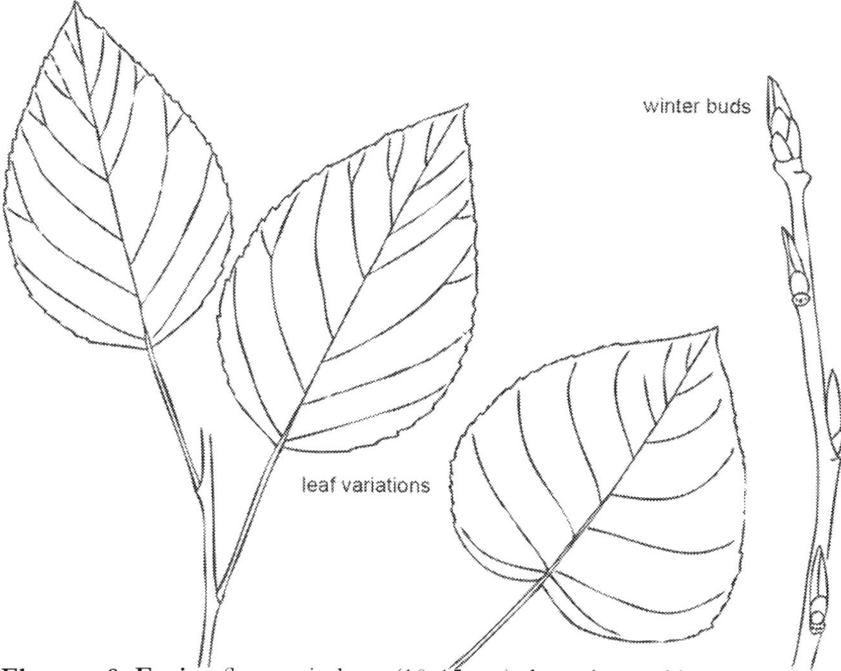

Flowers & Fruits: flowers in long (10-15 cm) drooping catkins, appearing before leaves; without calyx or corolla; fruit a capsule containing many small seeds with long white hairs
Leaves: alternate-simple, 7-12 cm; rounded, narrowing to sharp-pointed tip; base rounded to somewhat heart-shaped; lowest lateral veins not enlarged and do not turn parallel the edge of the leaf; leaves finely and evenly toothed or almost toothless, end of tooth rounded; leaf hairless except along the main veins on the underside; leaves dark green above, whitish waxy, often streaked with resin beneath; leaf stalk round or squarish; 2-3 glands may be present at the top of the leaf stalk
Buds & Twigs: buds 17-20 mm, smooth, gummy with spicy-resinous aroma; twigs round in cross-section, dark brown, hairy or smooth
Bark: light gray to; becoming dark and furrowed in older trees
Tree: to 20 m tall; diameter to 1 m
Habitat: wet woods, river banks, shores
Similar Species: Cottonwood, *P. deltoides*, and Quaking Aspen, *P. tremuloides*, have flattened leaf stalks

Eastern Cottonwood

Populus deltoides Bartram ex Marsh.

Status: FAC
Salicaceae

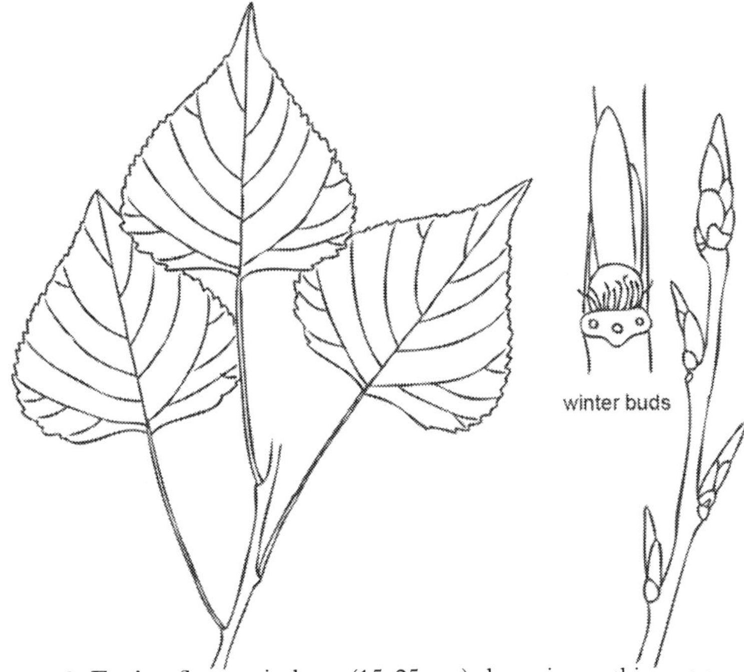

Flowers & Fruits: flowers in long (15-25 cm) drooping catkins, appearing before leaves; without calyx or corolla; fruit a capsule containing many small seeds with long white hairs
Leaves: alternate-simple, 8-14 cm; triangular, broad-based, narrowing to tip; base usually straight; lowest lateral veins not enlarged and do not turn parallel the edge of the leaf; leaves coarsely and evenly toothed, 15-25 teeth per side, teeth 1.5-2.0 mm deep, summit of tooth sharp; leaf hairless; edge of leaf with translucent green border; leaf stalk strongly flattened from side to side, 2-3 glands at the top of the leaf stalk
Buds & Twigs: buds ~ 20 mm, smooth, gummy but not fragrant; twigs round to 4-sided in cross-section, gray to reddish-brown, hairless
Bark: light gray, smooth; becoming dark and deeply furrowed in older trees
Tree: to 20 m tall; diameter to 1 m; tall with upper branches spreading forming a broad crown
Habitat: rich soils, floodplains, along stream banks
Similar Species: Quaking Aspen, *P. tremuloides*, also has flattened leaf stalks, but has glands on the leaf stalk at the base of the leaf

Quaking Aspen

Populus tremuloides Michx.

Status: Unknown
Salicaceae

Flowers & Fruits: flowers in long (10 cm) drooping catkins, appearing before leaves; without calyx or corolla; fruit a capsule containing many small seeds with long white hairs
Leaves: alternate-simple, rounded, base rounded to slightly heart-shape, narrowing to pointed tip; lowest lateral veins large and turn to parallel the edge of the leaf; leaves with small even teeth or almost toothless, 18-30 teeth per side, 1 mm deep, summit rounded; leaf hairless; leaf edge lacks translucent green border; leaf stalk strongly flattened from side to side
Buds & Twigs: buds 7-10 mm, smooth, not gummy and fragrant; edge of scales thin colorless; twigs round in cross-section, dark brown, hairless
Bark: light gray to gray-green; becoming dark and furrowed in older trees
Tree: to 15 m tall; diameter to 0.5 m; upper branches form rounded crown
Habitat: moist upland soils, streamsides, old fields
Similar Species: Bigtooth Aspen, *P. grandidentata*, is similar but each side of the leaf has 5-15 large teeth; Eastern Cottonwood, *P. deltoides*, also has flattened leaf stalks, but has a pair of glands on the leaf stalk at the base of the leaf

Black Cherry

Status: FACU

Prunus serotina Ehrh.

Rosaceae

Flowers & Fruit: flowers 5-parted with 1 pistil, ~ 20 stamens; petals ~ 4 mm; white; flowers in a loose raceme, 8-15 cm, at the end of the current year's twig; fruit in loose elongate grape-like clusters, dark-purple to black, 1 cm, with a single stone; sepals relatively large compared to fruit

Leaves: alternate-simple, 6-12 cm, sharp-toothed, lance-shaped with a sharp tip; underside of midvein with whitish (new leaves) to reddish-brown hairs (unique among Cherries); leaf stalks often with paired glands near the base of the leaf

Buds & Twigs: bud scales pointed; broken twigs with a distinctive odor partially due to hydrocyanic acid

Bark: red-brown with white transverse lenticils; older bark dark, cracking into small irregular plates exposing reddish-brown beneath

Tree: to 25 m height; diameter to 1 m

Habitat: old forests, young woods, thickets, roadsides

Similar Species: this is the only species with the hairs along the midvein on the underside of the leaf; Fire Cherry, *P. pensylvanica*, has a similar bark but the flowers/fruits are in umbel-like clusters

Choke Cherry

Prunus virginiana L.

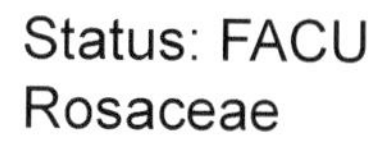
Status: FACU
Rosaceae

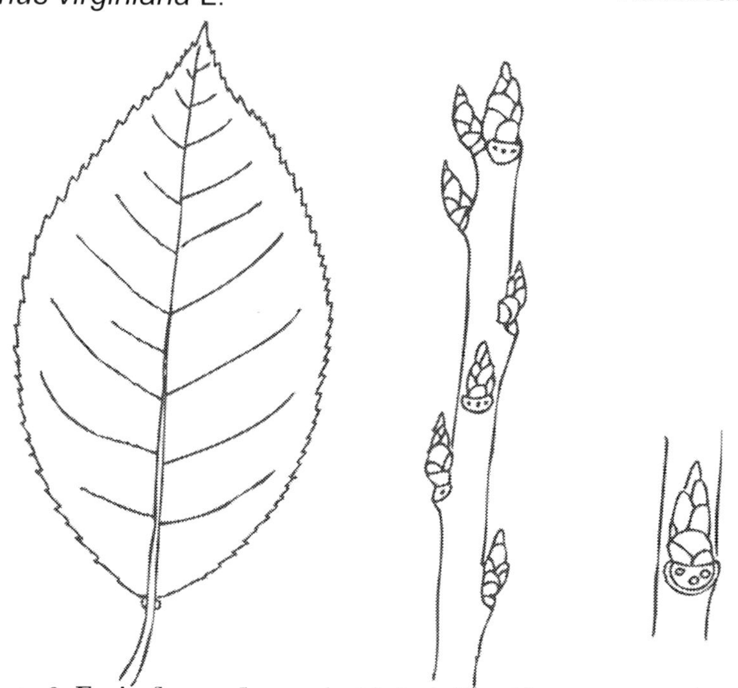

Flowers & Fruit: flowers 5-parted with 1 pistil, ~ 20 stamens; petals ~ 4 mm; white; flowers in a compact raceme, 6-15 cm, at the end of the current year's twig; fruit in elongate grape-like clusters, dark-red to black, 1 cm, with a single stone; astringent but edible; sepals relatively small compared to fruit
Leaves: alternate-simple, 5-12 cm, blunt-toothed, lance-shaped with a sharp tip; leaf stalks often with paired glands near the base of the leaf
Buds & Twigs: bud scales rounded; broken twigs with a distinctive odor partially due to hydrocyanic acid
Bark: smooth gray-brown
Tree: to 6 m height; diameter to 15 cm
Habitat: wide variety from hills and dunes to edges of swamps, young woods, thickets, roadsides
Similar Species: Black Cherry, *P. serotina,* has narrower leaves with brownish hairs on the underside of the midvein, reddish-brown bark, and flowers not as densely packed on the raceme; Fire Cherry *P. pensylvanicus,* has narrower leaves, reddish-brown bark, and the flowers and fruits are in a rounded cluster, each with only a few flowers

Swamp White Oak

Quercus bicolor Willd.

Status: FACW+
Fagaceae

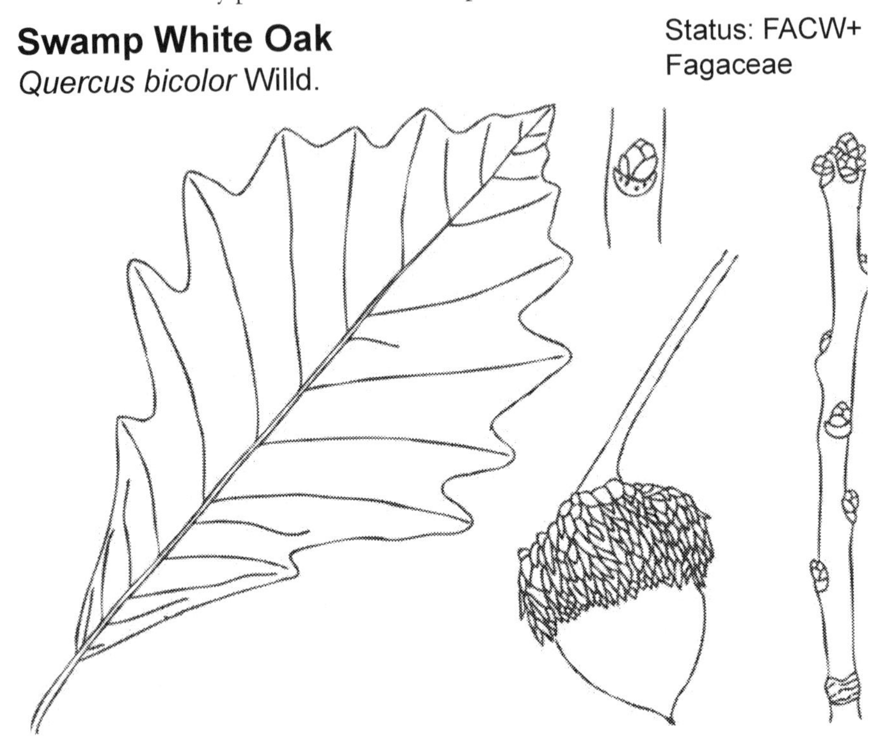

Fruits: acorns often paired, long-stalked (4-7 cm, longer than the leaf stalk); cup bowl-shaped, hairy, scales near rim with short irregular awns, cup covers up to 1/2 of acorn
Leaves: variable shape, 12-18 cm long; typically with 6-10 pairs of large irregular rounded teeth or shallow lobes (or sometimes (especially in our area) lobed proximally and toothed distally); lobes without bristle tips; white felty upright hairs beneath
Twigs & Buds: twigs hairless; end buds less than 5 mm long, blunt, rounded in cross-section, hairless, chestnut-brown
Bark: dark, flat-ridged and deeply furrowed or flaky
Tree: to 30 m; diameter to 1 m
Habitat: flood plains or poorly drained ground
Similar Species: among the oaks with round-tipped leaves it is most likely to be confused with Bur Oak, *Q. macrocarpa,* which has leaves with deeper sinuses (greater than one-third of the distance to the midrib) on the outer end and elaborate fringe on the scales of the acorn cap

Pin Oak

Quercus palustris Munchh.

Status: FACW
Fagaceae

Flowers & Fruits: male flowers are yellow-green catkins; reddish green female flowers on short spikes; round acorns, 12 mm long, with saucer-like cap with reddish brown scales covering one-quarter to one-third of nut
Leaves: alternate, simple, 5-20 cm long; 5-12 cm wide, oval with 5-9 deep cut (more than two-thirds of the distance to the midvein) bristle-tipped lobes, major lobes form a U-shape and widen at the outer ends: bright green above and pale green below; russet-bronze to red in fall
Buds & Twigs: slender, thin, greenish to reddish brown with small reddish brown terminal buds
Bark: grayish brown, smooth when young, with shallow furrows and thin ridges when older
Tree: up to 30 m tall, 1 m in diameter
Habitat: low, periodically flooded areas
Similar Species: U-shaped lobes distinguishes this species from other Adirondack oak species with bristle-tipped leaves

Alderleaf Buckthorn

Rhamnus alnifolia L'Her.

Status: OBL
Rhamnaceae

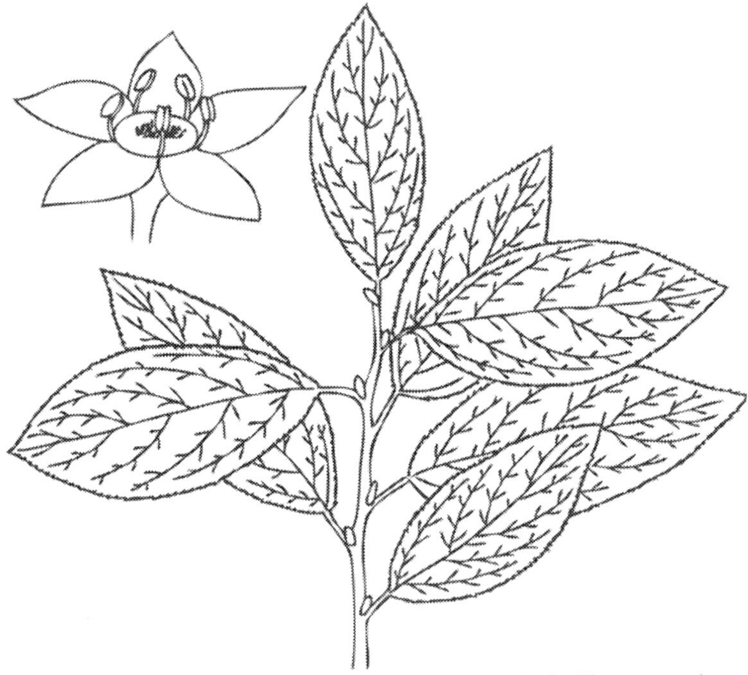

Flowers & Fruits: plants unisexual (stamenate and pistillate parts in separate flowers); 5-parted, trangular shaped sepals (no petals) 25-80 mm long; yellowish green to brown; in small clusters; fruit black berries with three seeds

Leaves: alternate, simple, yellowish green to green, up to 10 cm long and 5 cm wide; elliptic to oval, finely toothed, smooth or slightly hairy; 6-8 paris of veins which follow the leaf edges

Buds & Twigs: buds with dark scales; twigs reddish, hairless

Shrub: up to 1 m tall

Habitat: bogs, fens, swamps, wet meadows

Similar Species: similar to European Alderleaf Buckthorn, *R. frangula*, but that species has flowers with both sexes, fruits with two seeds, and leaves that are hairy underneath; see also Common Buckthorn, *R. catharctica*

Willows, *Salix spp*

A widespread genus of shrubs and small trees. All members of the genus have alternate-simple leaves with 3 vascular bundles in the leaf and single scales covering buds. The leaves are often long and narrow and often have teeth. Flowering occurs in early spring and the flowers are in long catkins that later develop the downy wind-blown seeds.

The willows are often difficult to identify because the vegetative characters vary a lot and often overlap; identification is further complicated by the existence of hybrids and polyploid clones.

The flow chart contains the species known from the Adirondacks. It should help narrow the number of species you need to consider:

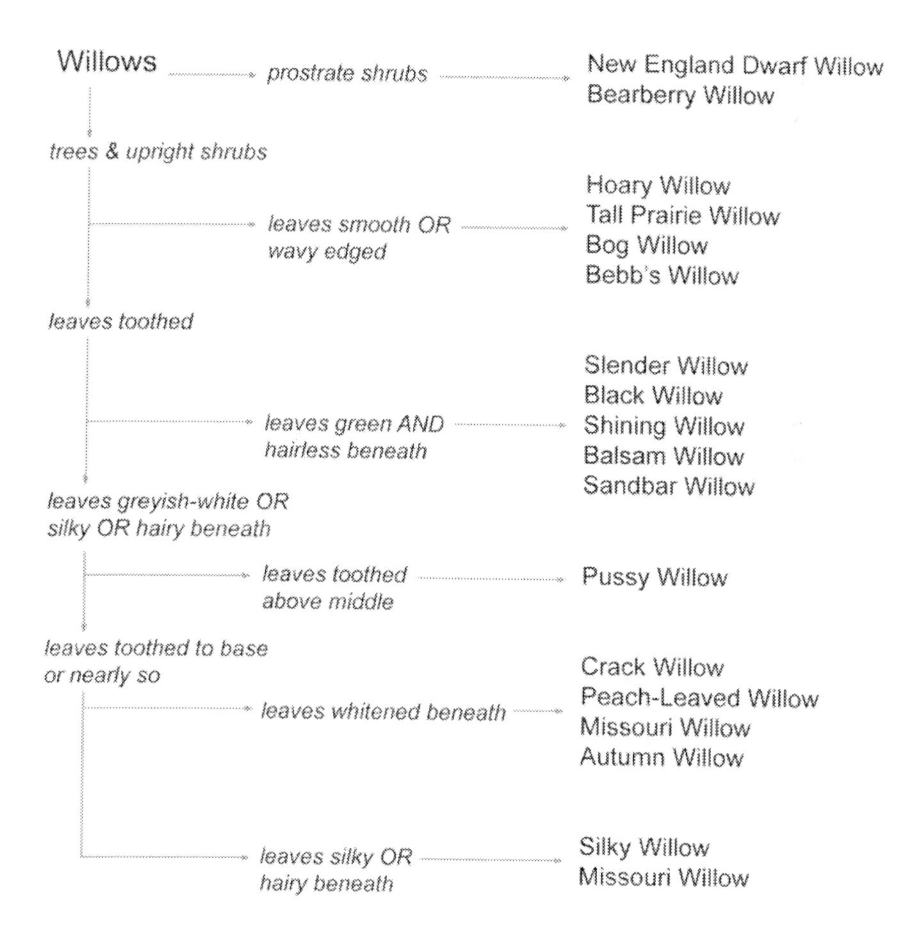

Bearberry Willow

Salix uva-ursi Pursh

Status: Unknown
Salicaceae

Flowers & Fruits: inflorescence a catkin; male catkins, 6-17 mm by 5-8 mm; female catkins, 8-39 mm by 6-10 mm
Leaves: shiny green, obovate to oblanceolate, flat, 8-23 mm by 3-10 mm, leathery, sharp fine-toothed margins
Buds & Twigs: reddish brown to grayish brown to yellowish brown, hairless
Shrub: matting, above ground brancehes do not take root
Habitat: wet alpine tundra and rockfaces
Similar Species: New England Dwarf Willow (aka Wideleaf Dwarf Willow), *S. herbacea*, is also a mat forming alpine species; the leaves are somewhat wider, heart-shaped at the base, and have obviously netted veins beneath; the above ground branches take root

Hoary Willow

Status: OBL

Salix candida Fluegge ex Willd.

Salicaceae

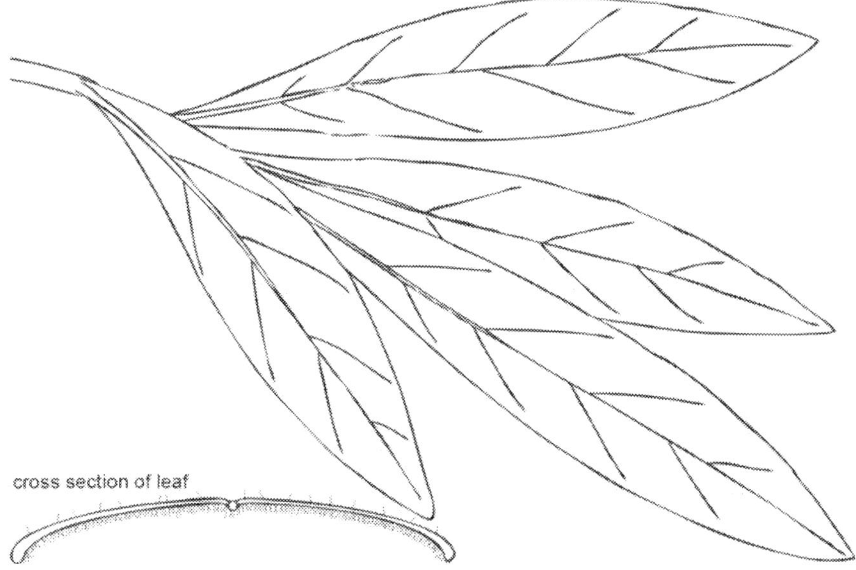

Flowers & Fruits: catkins on long shoots (0.5-2.0 cm) with bract like leaves, hairy, mature as leaves appear; catkins 1-5 cm; bracts brown with white hairs; fruits capsules cylindrical, 5-8 mm long, hairy
Leaves: alternate-simple; 4-8 cm; length ~ 5-6 x width; oblong to linear lanceolate; tip acute; teeth none to many, regularly spaced, rounded tips; margin rolled under; base tapered; upperside dull green with sparse hairs and sunken veins; underside whitish from a dense hairy coat that can often be rubbed off with the finger; petiole glands absent; stipules lanceolate, glandular
Buds & Twigs: scales fused; twigs with dense white wooly hair
Shrub: low, branched shrub; 0.5-2.0 m
Habitat: calcareous wetlands in glaciated areas
Similar Species: the long leaves with the margins rolled under and a dense white hairy undercoat that can often be rubbed off with your finger are the best way to distinguish this species

Tall Prairie Willow

Salix humilis Marsh.

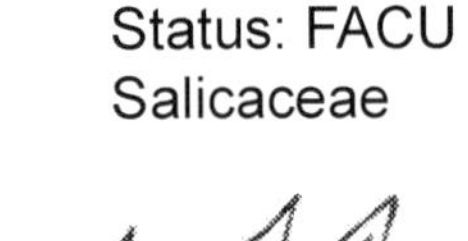

Status: FACU
Salicaceae

Flowers & Fruits: staminate catkins 1.5-4.0 cm; pistillate catkins 1.0-2.0 cm, developing small cottony seeds
Leaves: obovate to obovate-lance, base tapered, 4-8 cm by 1-2.5 cm; wavy edged, in-rolled margins; dark green above, grayish-white hairy beneath, with yellow veins
Buds & Twigs: yellowish brown to dark brown, usually gray hairy
Shrub: up to 3 m tall
Habitat: open woods, wet prairies
Similar Species: the grayish hairs on the undersides of the leaves distinguish this from other species in the smooth-wavy edged leaf group

Bog Willow

Salix pedicellaris Pursh.

Status: OBL
Salicaceae

Flowers & Fruits: green to brown catkins; oval capsule with wooly seeds
Leaves: alternate, simple; elliptic to obovate, 1-5 cm long, smooth margins, green above, pale green to bluish green underneath; leathery, hairless
Shrub: clumping, up to 1 m tall
Habitat: bogs, fens, swamps
Similar Species: the hairless leaves distinguish this from the other species in the smooth-wavy edged group

Bebb Willow or Diamondbark Willow

Status: FACW

Salix bebbiana Sarg.

Salicaceae

Flowers & Fruits: catkins on short leafy shoots, appear before leaves; staminate catkins 1-2 cm, pistillate catkins 2-6 cm; fruit capsules 6-8 mm, stalks 6-8 mm, long-beaked, sparsely hairy

Leaves: alternate-simple; 4-8 cm, length 2.5 x width; elliptical to oval, relatively wide; tip acute to acuminate; teeth few to many, regularly spaced, rounded tips; margin flat; base tapered; upperside gray-green, usually with some gray to white hairs; undersides whitish waxy sheen with dense gray to white wooly hairs; petiole glands absent; stipules absent to minute

Buds & Twigs: buds blunt, shiny brown; scales fused; twigs reddish-purple to orange-brown; branchlets hairy when young, then hairless; stipules none

Bark: ridges often form diamond shape

Shrub: tall shrub to small tree, 2-5 m

Habitat: moist or wet places

Similar Species: the large, almost tree-like, size and relatively wide leaves help distinguish this species from all others except Pussy Willow, *S. discolor*, and Bog Willow, *S. pedicellaris*, from which it differs by having the diamond shape ridges in the bark and not having conspicuous stipules

Slender Willow or Meadow Willow

Status: OBL

Salix petiolaris Sm.

Salicaceae

Flowers & Fruits: catkins appear with leaves on short leafy shoots; pistillate catkins to 4 cm; fruit capsules 3-7 mm, elongated with slender beaks, on slender hairy stalks
Leaves: alternate-simple, 4-10 cm, length 5 x width or more; narrow-lanceolate, sides almost parallel much of length; tip acute to acuminate; teeth few to 5 per cm, regularly spaced, sharp tips with glands; margin flat; base tapered; upperside dark green; underside variably whitish waxy sheen when mature, silky hairs when young; petiole glands absent; stipule small to absent
Buds & Twigs: bud scales fused; twigs slender, yellow-hairy to brown-hairless; branchlets brittle at base
Bark: gray-green to red-brown, becoming dark brown with age
Shrub: clumpy shrub to multi-trunked tree, to 7 m
Habitat: moist meadows, streambanks, lake shores
Similar Species: the combination of small to absent stipules, leaf undersides hairless, leaf tip not elongate, and being a shrub distinguish this species from others

Black Willow

Salix nigra Marsh.

Status: FACW+
Salicaceae

Flowers & Fruits: catkins erect, 2-7 cm, on short leafy shoots; appear in early spring; bracts 2-3 mm, blunt tipped, yellow, hairy on the inside; fruit capsules ovoid, 4-5 mm, light brown, hairless
Leaves: alternate-simple; 6-10 cm, length 5-6 x width; shape lanceolate, narrow, often droop on leaf stalk; tip acute to acuminate, often sickle shape near tip; teeth at least 5 per cm, regularly spaced, sharp teeth; margin flat, base rounded; upperside deep green, veins unite and run along the margin of the leaf; underside paler green, hairless to slightly silky; petiole glands absent; stipules heart-shaped, acute, toothed, to 12 mm
Buds & Twigs: buds pointed, 3-4 mm, scales not fused, yellowish-brown, shiny; twigs light yellow to red/purple-brown, ridges run down twigs from leaf scars; branchlets brittle at base, no waxy bloom
Bark: dark brown to blackish; deeply furrowed, scaly flat-topped ridges
Tree: shrub to tree, often with 1-4 leaning trunks
Habitat: alluvial soil along streams and in wet meadows
Similar Species: Crack Willow, *S. fragilis,* the other tree-sized species, has much smaller leaf stipules

Shining Willow
Salix lucida Muhl.

Status: OBL
Salicaceae

Flowers & Fruits: catkins appear with leaves on short leafy shoots; staminate catkins 2-4 cm; pistillate catkins 2-5 cm; fruit capsules 4-7 mm, elongated with slender beaks, on slender hairy stalks
Leaves: alternate-simple, 5-15 cm, length 5 x width or more; narrow-lanceolate, sides almost parallel part of length; tip acute to acuminate; teeth at least 5 per cm, regularly spaced, sharp tips with glands; margin flat; base tapered; upperside dark lustrous green; underside variably whitish waxy sheen when mature, silky hairs when very young (soon lost); petiole glands absent; stipule small to absent, egg-shaped to crescent-shape, up to 5 mm
Buds & Twigs: bud scales fused; twigs slender, yellow-hairy to brown-hairless; branchlets brittle at base
Bark: gray-green to red-brown, becoming dark brown with age
Shrub: clumpy shrub to multi-trunked tree, to 5 m
Habitat: moist meadows, streambanks, lake shores
Similar Species: Balsam Willow, *S. pyrifolia*, is similar but has leaves that are aromatic when crushed; the combination of small to absent stipules, shining upper leaf surface, leaf undersides hairless, leaf tip elongate distinguish Shining Willow from others in the genus

Sandbar Willow

Salix interior Rowlee

Status: Unknown
Salicaceae

Flowers & Fruits: catkins 1-3 on short lateral spike in axils of previous year's buds; pistillate catkins 2-6 cm
Leaves: alternate, linear-oblong, 5-14 cm long, acute, teeth widely spaced along margin; green above, paler beneath; whitish green petioles, 3 mm long; stipules none
Buds & Twigs: green to reddish green, with spongy white pith
Shrub: shrubs to multi-trunked tree up to 5 m tall
Habitat: swamps, streambanks, flood plains
Similar Species: long slender hairless leaves with widely spaced teeth distinguish this species from other species in the group
aka: *S. exigua*

Pussy Willow

Salix discolor Muhl.

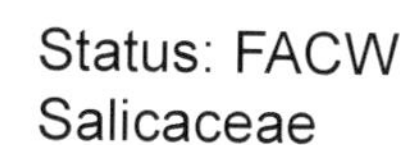

Flowers & Fruits: catkins on short shoots with bract like leaves; densely covered by silky hairs when immature, mature before leaves appear; staminate catkins 2-4 cm; pistillate catkins 2-6 cm; bracts dark brown with white hairs; fruit capsules cylindrical, 7-12 mm, long-beaked, hairy

Leaves: alternate-simple; 4-8 cm; length ~ 2.5 x width; elliptical to oblong, relatively wide, base tapered, tip acute to acuminate; teeth few to many, above middle of leaf, rounded tips; margin flat; upperside bright green with raised veins; underside whitish green hairless to rusty-hairy (especially when young); petiole glands absent, stipules small to large, elliptical to semi-oval

Buds & Twigs: buds large (to 7 mm), flattened, pointed, reddish-purple, scales fused; twigs stout, reddish to brownish, hairy when young; branchlets without waxy bloom

Bark: grayish-brown, not furrowed

Shrub: tall shrub to small tree; 2-5 m

Habitat: wet meadows, swamps, shorelines

Similar Species: the only species with leaves consistently toothed only above the middle

Crack Willow

Salix fragilis L.

Status: FAC+
Salicaceae

Flowers & Fruits: catkins on leafy shoots (1-3 cm) with 2-5 leaves, mature as leaves appear; catkins 4-8 cm; bracts greenish-yellow, deciduous; fruits capsules narrow-conical, 4.0-5.5 mm long, hairless
Leaves: alternate-simple; 7-12 cm; length 3-4 x width; lanceolate; tip acute to acuminate; teeth five or more per cm, regularly spaced, sharp tips with glands at the tip; margin flat; base tapered; upperside dark green; underside green, hairless, variable waxy sheen; petiole glands small on upper surface or absent; stipules small, lance-shaped
Buds & Twigs: buds gummy, scales fused; twigs stout, greenish-yellow to dark-red; branchlets brittle at base; no waxy bloom
Bark: deeply furrowed, ridges narrow
Shrub: tree, to 20 m tall; diameter to 1 m
Habitat: wet meadows, swamps, shorelines
Similar Species: the only other tree-size willow is Black Willow, *S. nigra,* which has much larger heart-shaped stipules

Missouri Willow

Salix eriocephala Michx.

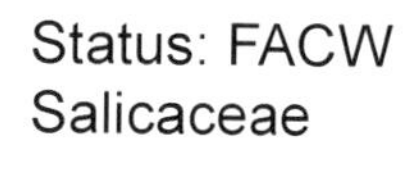
Status: FACW
Salicaceae

Flowers & Fruits: erect catkins, 2-5 cm, appear before leaves
Leaves: aternate, simple; oblong to lance-like, toothed margin, base rounded to heart-shaped; long petioles, 5-14 mm; dark green above, whitened or hairy below; stipules leafy, 5-10 mm
Buds & Twigs: buds reddish-brown
Shrub: up to 5 m tall
Habitat: streambanks, lowlands
Similar Species: The leaves being heart-shaped at the base distinguishes this species from Peach-Leaved Willow, *S. amygdaloides*, which has leaves with tapered bases

Autumn Willow

Salix serrissima (L.H. Bailey) Fernald

Status: OBL
Salicaceae

Flowers & Fruits: catkins 2-4 cm long, on short leafy stalks
Leaves: alternate, elliptic, lance-like to oblong, 5-10 cm long and 1.0-2.5 cm wide, tapered tips, toothed margin, shiny dark green above, whitened underside; stipules minute or none
Buds & Twigs: reddish brown
Bark: grayish brown
Shrub: up to 3 m tall
Habitat: bogs, swamps, wet meadows, lakeshores, streambanks
Similar Species: the lack of leafy stipules along with being a shrub rather than a tree help distinguish this species from others in the group

Silky Willow

Salix sericea Marsh.

Status: OBL
Salicaceae

Flowers & Fruits: catkins 1-4 cm, stalkless or stalk less than 1 cm, appear before leaves; scales 1 mm, blackish; fruits ovoid, 3-5 mm
Leaves: alternate-simple; 6-10 cm, length 4-6 x width; shape narrow lanceolate, tip acute to acuminate; teeth at least 5 per cm, regularly spaced, sharp tips bearing glands; margin flat; base wedge-shaped; upperside dark green; underside with dense silver-silky hairs (use lens); petiole glands absent; stipule lance-shaped to ovate, 3-10 mm, deciduous
Buds & Twigs: bud scales fused; twigs brittle at base, no waxy bloom; branchlets flexible at base
Shrub: shrub to small tree, to 4 m
Habitat: moist or rocky ground, often in or near running water
Similar Species: the dense silver-silky hairs on the underside of the leaf distinguish this species from others

Meadowsweet

Spiraea alba DuRoi

Status: FACW+
Rosaceae

Flowers & Fruits: flowers small (3-4 mm), 5-parted, white; in erect clusters; fruit small, dry, hairy
Leaves: alternate-simple, 3-7 cm; elliptical, fine-toothed, relatively narrow (2-4 times as long as wide), hairless underneath
Buds & Twigs: buds long-pointed, silky; twigs yellow-brown to reddish-brown
Bark: papery
Shrub: to 2 m
Habitat: wet meadows, swamps, shores
Similar Species: sometimes divided into variety *alba* (leaf length 3-4 times as long as wide, yellowish-brown twigs) and variety *latifolia* (leaf length 2-3 times as long as wide, reddish-brown twigs); Steeplebush, *S. tomentosa,* has wooly hairs on the twigs and undersides of leaves

Steeplebush

Spiraea tomentosa L.

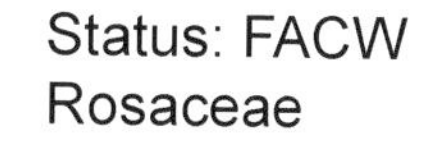
Status: FACW
Rosaceae

Flowers & Fruits: flowers small (3-4 mm), 5-parted, pinkish; in long erect clusters; fruit small, dry, hairy
Leaves: alternate-simple, 3-5 cm; egg-shaped to oblong, coarse-toothed, undersides covered with dense whitish or rusty wooly hairs
Buds & Twigs: buds long-pointed, silky; twigs wooly
Bark: papery
Shrub: to 1.2 m
Habitat: moist meadows, swamps
Similar Species: Meadowsweet, *S. alba,* does not have wooly twigs or hair on the undersides of leaves

American Elm

Status: FACW-

Ulmus americana L.

Ulmaceae

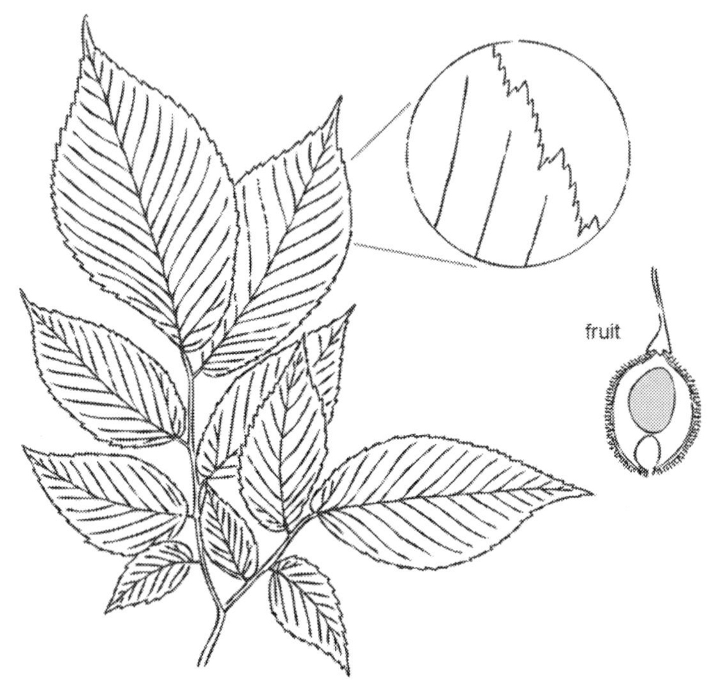

Flowers & Fruits: flowers in short, loose tassel-like clusters; fruits round, flat, winged, ~ 1 cm diameter; hairy only along margins of wing
Leaves: alternate-simple, 8-14 cm; variable; egg-shaped with pointed tip, double-toothed, bases uneven; upper surface dark-green, sometimes glossy, smooth to somewhat rough; lower surface hairy or smooth
Buds & Twigs: bud scales over 7 mm long, red-brown with dark edges; twigs hairless
Bark: young bark dark brown with shallow intersecting furrows; older bark deeply furrowed
Tree: trunk branches into large upturned limbs near ground giving the tree a unique vase-shape; to 30 m tall, diameter to 1.5 m
Habitat: bottomlands, moist, fertile soil
Similar Species: Slippery Elm, *U. rubra,* has leaves which are very rough on the upper surface; fruits are larger

Slippery Elm

Ulmus rubra Muhl.

Status: FAC
Ulmaceae

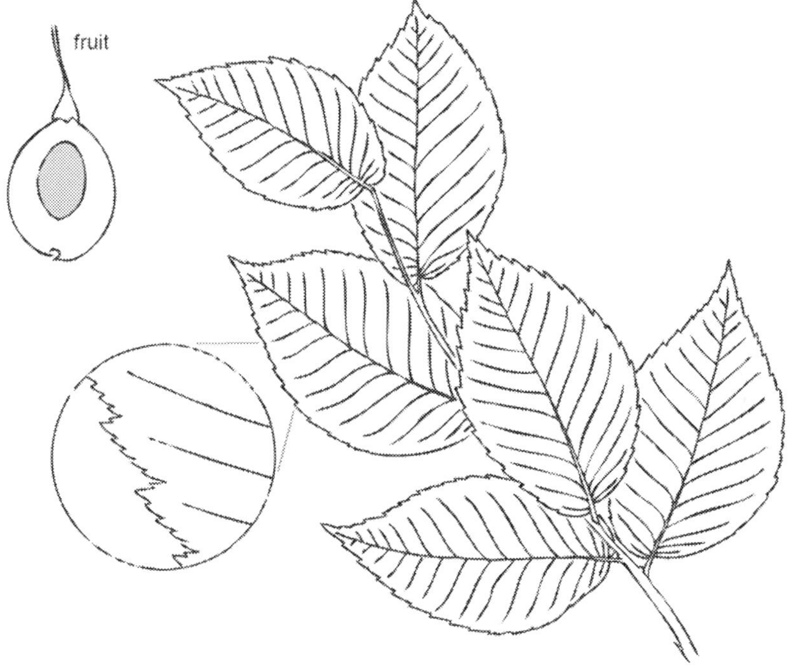

Flowers & Fruits: flowers in short, loose tassel-like clusters: fruits are round and papery, 1.5-2.0 cm, with smooth winged margin, hairy only on the seed capsule
Leaves: alternate-simple, 8-14 cm; egg-shaped with pointed tip, double-toothed, bases uneven; upper surface dark green, very rough; lower surface pale green, hairy
Buds & Twigs: twigs grayish-brown, hairy, slightly zigzag; with greyish-black lateral buds, hairy, bud scales over 3 mm
Bark: reddish brown to grayish brown, deeply furrowed, fibrous
Tree: to 18 m tall and 50 cm in diameter; branches upturned, giving tree characteristic elm vase-shape outline
Habitat: swamps, moist woods and fields
Similar Species: American Elm, *U. americana*, has leaves with a smoother, sometimes glossy, upper surface

Grass-like Plants

The "grass-like" or "graminoid" plants are usually more difficult to identify than the other groups you will encounter. Although the structure of the flowers is important for identification, it is highly modified in most of these groups. Also, the structures are usually small and require either a hand lens or, better yet, a dissecting microscope.

Many of these groups contain large numbers of species. Identification in these groups should always be tentative and you should always bring a sample back to allow confirmation of your field identification.

Although they are difficut to identify, the grass-like plants are nevertheless important components of wetlands and cannot be ignored. We have tried to make the guide complete enough that you can at least identify the group to which your specimen belongs, and in some cases the species. However, you will need to refer to specialized works to confirm or refine your identifications.

We have included three families in the "grass-like" plants. Each family is described below and this, along with browsing the illustrations should allow you to place your specimen in the proper family.

Rushes (Juncaceae): These are somewhat grass-like in general appearance, but the flowers when viewed close-up, are similar to traditinal flowers: they are definitely 3-parted, with a 3-parted style, 3 or 6 stamens, 3 small bract-like petals and 3 small bract-like sepals. Later in the season a capsule-like fruit develops but the remnants of the petals and sepals remain.
Sedges (Cyperaceae): The flowers lack both petals and sepasls. There is a usually a single bract arising just below the flower. The flowers are often unisexual and the different sexed flowers located on different spikes or on different parts of the same spike. The stems, especially near the base, are often triangular in cross-section and are usually solid. The leaves are arranged on the stems in three rows. Except in the genus Carex there is no ligule where the blades come off of the stem and the leaf sheaths are "closed".
True Grasses (Poaceae): The flowers lack both petals and sepals but have two bracts, which arise on alternate sides just below the flower. The flowers are usually bisexual, so sexually differentiated spikes do not usually occur the way they do in the sedge family. The stems, especially near the base, are round in cross-section and are hollow. The leaves are arranged on the stems in two rows. There is a ligule where the blade comes off the stem and the sheaths are "open".

Rushes (Juncaceae):

As described earlier the rushes are somewhat grass-like in general appearance, but the flowers when viewed close-up, are similar to traditional flowers. The flowers are 3-parted, with a 3-parted ovary a style that ends in 3 stigmas, either 3 or 6 stamens, 3 small bract-like petals and 3 small bract-like sepals (sepals + petals = tepals). Later in the season a capsule-like fruit develops but the remnants of the petals and sepals remain. At maturity the capsule splits open into three sections.

There are two genera in the Adirondacks: *Juncus* (24 species) and *Luzula* (5 species). In *Juncus* the capsule has several seeds and the leaf margins are hairless; *Luzula* is very similar to Juncus but has only 3 seeds in the capsule and the leaf-margins have long hairs along the edges. In the following pages 6 different *Juncus* are figured.

A structure someimes found in rushes that you will need to use to identify them is the prophyll. Each flower is set on the end of a stalk (the pedicel); just under the sepals there may be a set of two small bracts, each called a prophyll. Plants with prophylls are "prophyllate"; those without are "eprophyllate".

There are only a few species of prophyllate *Juncus* in the Adirondacks, but they are among the ones more commonly found: Toad Rush, *J. bufonius*; Soft Rush, *J. effusus*; Thread Rush, *J. filiformis;* Saltmeadow Rush, *J. gerardii* and Path Rush, *J. tenuis.*

The eprophyllate species are Sharp-Fruit, *J. acuminatus*; Alpine Rush, *J. alpinoarticulatus*; Jointed Rush, *J. articulatus*; Small-Head Rush, *J. brachycephalus*; Narrow-Panicle Rush, *J. brevicaudatus*; Canada Rush, *J. canadensis*; Grass-Leaved Rush, *J. marginatus*; Bayonet Rush, *J. militaris*; Knotted Rush, *J. nodosus*; Brown-Fruited Rush, *J. pelocarpus*; Needle-Pod Rush, *J. scirpoides*; American Moor Rush, *J. stygius*; Woods Rush, *J. subcaudatus*; and Torrey's Rush, *J. torreyi.*

graminoids: rushes

Toad Rush

Juncus bufonius L.

Status: FACW
Juncaceae

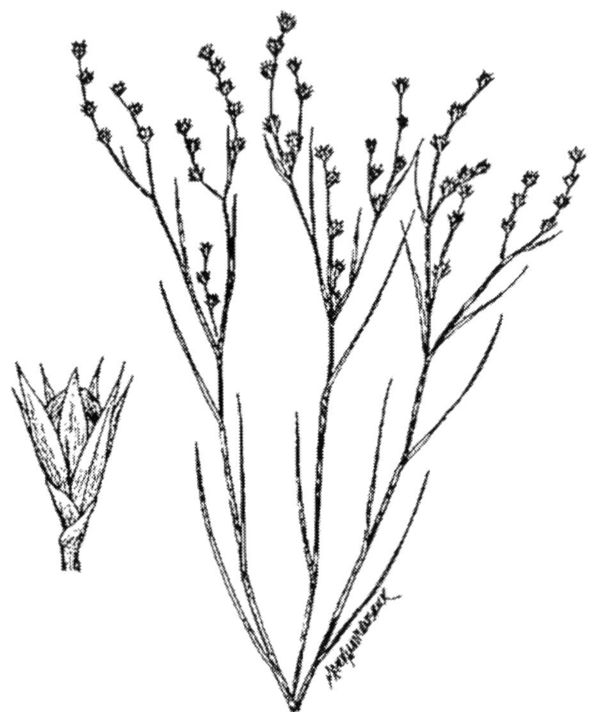

Plants: up to 30 cm tall; annual
Leaves: both basal and cauline present; 3-13 cm long by 0.3-1.1 mm wide
Infloresence: loose and diffuse; infloresence more than one-third the total height of the plant; only short bract or none at base of infloresence
Flower: prophyllate, tepals slender-lanceolate, 3-6 mm, with greenish midstripe, outer a little longer than inner
Achenes: 0.3-0.6 mm
Habitat: wet meadows, shorelines, streambanks, roadside
Similar Species: Saltmeadow Rush, *J. gerardii*, is the only other prophyllate species that has both basal and cauline leaves but it has an infloresence no more than one-quarter the total height of the plant

Soft Rush

Juncus effusus L.

Status: FACW+
Juncaceae

Plant: forms dense, often large, clumps
Stems: erect, 40-130 cm tall by 1.0-2.5 mm in diameter at tops of sheaths; round, may have tiny ridges
Leaves: bases of stems have open reddish-brown sheaths up to 20 cm long but no blade
Infloresences: several lateral compound; primary bract extends beyond infloresence appearing to be a continuation of the stem so the infloresence appears to be lateral; bract at least one-quarter the length of the rest of the stem
Flowers: prophyllate; 3 sepals; 3 slightly shorter petals; usually tan with greenish midstripe; 3 stamens
Fruit: capsule 1.5-3.2 mm tall
Seeds: amber, 0.4-0.5 mm
Habitat: swamps, marshes, moist or saturated soils, especially in pastures
Similar Species: Thread Rush, *J. filiformis*, is similar but the stem is only 1.0 mm in diameter and the bract is one-half to at least as long as the rest of the stem; 6 stamens

Saltmeadow Rush

Juncus gerardii Loisel.

Status: FACW+
Juncaceae

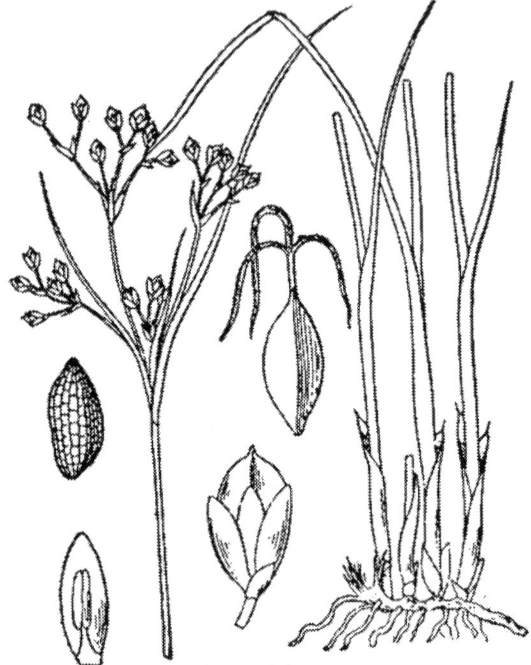

Plants: 20-60 cm tall; perennial; from rhizome
Leaves: basal and 1-2 cauline leaves, the uppermost diverges from the stem ~ one-half way up; flat or channeled, 10-40 cm by 0.4-0.7 mm
Infloresence: terminal; 10-30 flower heads, loose; 2-8 cm long; bract on the infloresence is shorter than the infloresence (notice that in the illustration above this character is not shown correctly)
Flower: prophyllate, tepals dark brown to blackish with greenish midstripe, lanceolate-ovate to oblong; 2.4-3.2 mm long
Achenes: dark brown, 0.5-0.7 mm
Habitat: edge of ponds, marshes, disturbed areas
Similar Species: Toad Rush, *J. bufonius*, is the only other prophyllate species that has both basal and cauline leaves but it has an infloresence more than one-third the total height of the plant; another common prophyllate species is the Path Rush, *J. tenuis*, which has basal leaves but lacks stem leaves and the bract on the infloresence is longer than the infloresence itself

Tapertip Rush

Juncus acuminatus Michx.

Status: OBL
Juncaceae

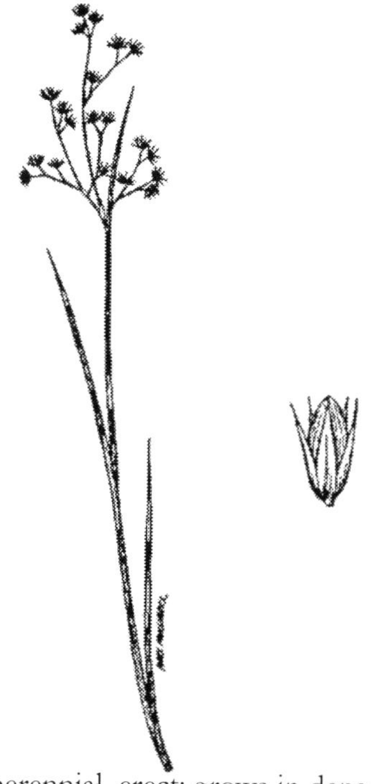

Plants: 20-80 cm tall, perennial, erect; grows in dense tufts
Stems: angled
Leaves: 1-2 basal and 1-2 cauline; sharply angled, 1-40 cm long, with internal partitions 1-3 mm thick
Infloresence: at ends of branched panicle, clusters of 5-20 (50) heads, 5-15 cm long,
Flower: eprophyllate; light brown to greenish, lanceolate, 2.6-3.9 mm long; sepals with chaffy margins; 3(6) stamens
Habitat: wet soils of marshes, meadows, shores, low woods
Similar Species: blades of leaves of all eprophyllate species except Moor Rush, *J. stygius*, Grass-Leaved Rush, *J. marginatus*, and, sometimes, Brown-Fruited Rush, *J. pelocarpus*, have internal partitions (septae); identification of most of the others depends on characters such as seed size, structure of the fruit capsule, and number of stamens; the next two species are distinctive enough that they can be identified without a taxonomic key

Bayonet Rush

Juncus militaris Bigelow

Status: OBL
Juncaceae

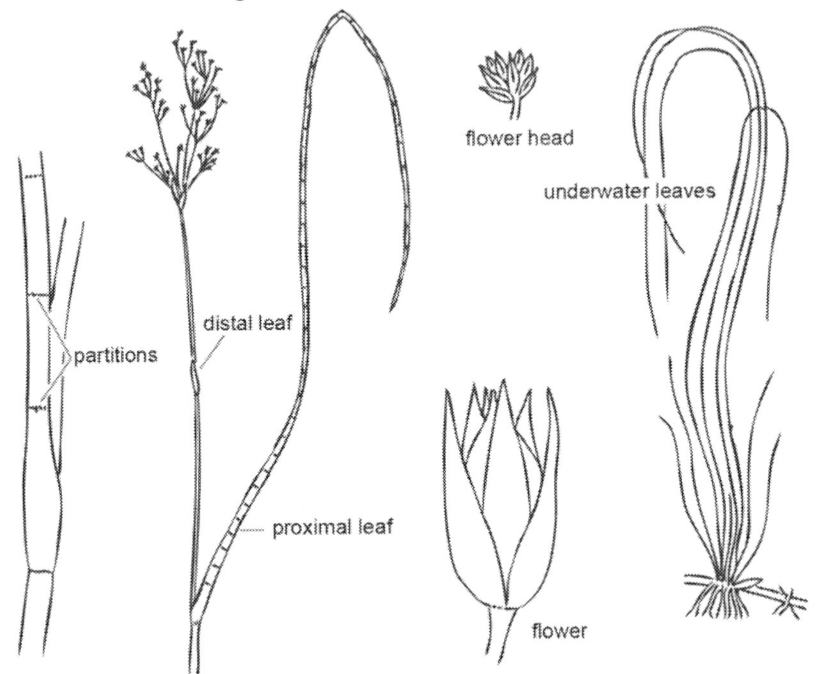

Plants: perennial; from rhizome
Stems: erect, 50-100 cm; 5-12 mm diameter
Leaves: 1-2 basal bladeless leaves; 1 (2) emergent cauline leaves diverging near middle of stem, blade stout, overtopping infloresence, with hard cross-partitions; in running water rhizome may produce long hair-like leaves
Infloresence: 4-15 cm; ascending branches ending in 20-100 heads, each with 5-13 (25) flowers
Flower: eprophyllate; pale brown to reddish brown; laceolate, 2.3-3.5 mm long
Achenes: yellowish brown, three-angled, obovoid, 2.4-3.2 mm with conspicuous beak
Habitat: shallow mucky waters
Similar Species: the habitat and large conspicuously septated leaf blade are distinctive

Brown-Fruited Rush

Juncus pelocarpus E. Mey.

Status: OBL
Juncaceae

Plants: erect, perennial, from rhizome
Leaves: 1-2 basel and 1-4 cauline; 1.5-11.0 cm long by 0.8-1.1 mm wide (narrow); blade angles, faint partitions or none
Infloresence: terminal, with central axis, 5-15 cm; single or double flowers clustered on one side of branches
Flower: eprophyllate; 1-2 (3) flowers in each head; pale brown, sepals 1.6-2.3 mm, petals 1.8-2.8 mm, oblong
Achenes: brown; 0.3-0.5 mm long
Habitat: sandy or boggy soil at edge of lakes and ponds, occasionally submerged
Similar Species: only eprophyllate species where the flowers are born in small heads of 1-2 (3) flowers rather than in larger cluster

Sedge Family (Cyperaceae): Many members of this family are adapted to wetlands in cool climates.

The flowers lack both petals and sepals. The flowers may be either bisexual or unisexual (with the different sexed flowers located on the same plant). There is a usually a single bract arising just below the flower. The flowers are arranged on a cental axis as spikes, which may have a bract as its base. The spikes are arranged into an infloresence in various ways.

The stems, especially near the base, are often triangular in cross-section and are usually solid. The leaves are either basal or along the stem or both. They are arranged in three rows, which can be most easily seen on the stem leaves. Except in the genus *Carex* there is no ligule where the blades come off of the stem and the leaf sheaths are "closed", i.e. the opposite edges of the sheath wrap around the stem and fuse together in the front where they meet.

In all the below except *Cypress* and *Dulichium* the scales in the spikes are arranged spirally, giving the spike a cylindrical rather than a flattened shape. In all except *Scleria* and *Carex* the flowers are unisexual and there are separate spiklets or regions of spikes for each sex.

There are fourteen genera in the Adirondacks (they are arranged in this order in the text):

Bulrush, *Scirpus* (7 species): sepals and petals reduced to bristles; branching infloresence at top of stem with bracts reflexed underneath; *(Go to page 120)*

Bulrush, *Schoenoplectus* (8 species): similar to Scirpus but a single bract extends beyond the infloresence appearing to be continuation of stem so the infloresence appears to be lateral; *(Go to page 123)*

Bulrush, *Bolboschoenus* (1 species): similar to *Scirpus* but with spikelets 6-10 mm in diameter rather than less than 3.5 mm in *Scirpus;* *(Go to page 127)*

Bulrush, *Trichophorum* (3 species): similar to *Scirpus* but with a single terminal spike of a few modified scales and a few flowers; Alpine Cotton-Grass, *T. alpinum*, Tufted Bulrush, *T. cespitosum*; and Clinton's Bulrush, *T. clintonii*

Cotton-Grass, *Eriophorum* (5 species): petals and sepals are much elongated at maturity so that mature spikes form a dense cottony head; *(Go to page 128)*

Spikerush, *Eleocharis* (10 species): leaves reduced to basal sheaths; solitary spikes terminal on stem with sterile scales at bottom; achene has a cap on top, the tubercule, formed from the remnants of the style; *(Go to page 129)*

Fimbry, *Fimbristylis* (1 species): similar to *Scirpus* but with narrower leaves; upper ends sheaths entire or with short fringe; Slender Fimbry, *F. autumnalis*
Hair-Sedge, *Bulbostylis* (1 species): similar to *Scirpus* but with narrow, wirey stems and basal leaves; upper ends of sheaths with long uneven fringe; achene with minute tubercule; upland; Densetuft Hairsedge, *B. capillaris*
Beak-Rush, *Rhynchospora* (4 species): similar to *Scirpus*, thin, wirey, leafy stems; achene with prominent tubercule; *(Go to page 133)*
Saw-Grass, *Cladium* (1 species): similar to *Rhynchospora* but with thicker stems; achene without prominent tubercule: *(Go to page 134)*
Flatsedge, *Cypress* (9 species): similar to *Scirpus* but because the scales are arranged in 2 ranks the clusters and spikes appear flat; *(Go to page 135)*
Three-Way Sedge, *Dulichium* (1 species): infloresence similar to *Cypress*, but in leaf axils rather than terminal; *(Go to page 136)*
Bulrush, *Lipocarpha* (1 species): similar to *Scirpus* but achenes subtended by two basal scales; stem leaves clustered near base; Dwarf Bulrush, *L. maculata*
Nut-Rush, *Scleria* (2 species): staminate and pistillate spikes separate as in *Carex*, but achenes not enclosed in perigynium; Whip Nutrush, *S. triglomerata*, has smooth achenes; Low Nutrush, *S. verticillata*, has rough achenes
Sedges, *Carex* (119 species): stems triangular, at least near base; staminate and pistillate flowers on different spikes or different sections of spike; spikes usually multiple arranged along stem; perigynium present; *(Go to page 137)*

Dark Green Bulrush

Scirpus atrovirens (Willd.)

Status: OBL
Cyperaceae

Plants: grows in dense upright clumps
Leaves: 6-12 per stem; proximal leaves with light brown sheaths and many more or less conspicuous septa; blades 20-54 cm by 7-17 mm
Infloresence: terminal, sometimes also one lateral from upper leaf axil; branches usually twice branched, ascending and/or divergent; bases of involucral bracts green, margins speckled red-brown or occasionally black
Spikelets: in dense clusters, each with up to 25 spikelets; sessile, 2-5 mm by 1.0-2.5 mm; scales dark brown with pale midribs, elliptical, 1.2-2.1 mm long with a sharp tip
Flowers: 6 perianth bristles persistent, about same length as achene; bristles thin-walled; enclosed within the scales
Achenes: light brown to whitish, elliptical, 3 sided or flattened
Habitat: marshes, moist meadows
Similar Species: *S. hattorianus* is very similar but basal leaf sheaths have only a few faint septae and perianth bristles slightly shorter than the achenes; *S. microcarpus* (aka *S. rubrotinctus*) has dark red bands on leaf sheath

Blackgirdle Bulrush

Scirpus atrocinctus Fern.

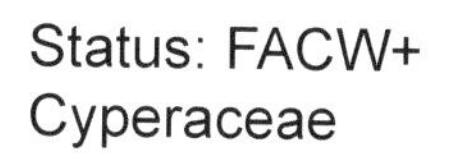
Status: FACW+
Cyperaceae

Plants: grows in dense clumps, with upright stem
Leaves: 4-7 per stem; blades 25-63 cm by 3-6 mm; proximal leaves with brown sheaths with few-many, +/- conspicuous septae
Infloresence: terminal; usually 2 x branched, branches ascending or spreading; long involucral bracts with black bases
Spikelets: in open cymes; spikelets mostly solitary with distinct pedicels, 4.0-7.0 mm by 2.0-2.7 mm; scales usually black, at least distally, oblong, 1.3-1.8 mm, with or without tiny tip
Flowers: 6 perianth bristles persistent, slender and much longer than achene projecting beyond scales, bristles smooth
Achenes: whitish, elliptical or obovate, 3-sided or flattened
Similar Species: the dark bases of involucral bract that create the appearance of a black girdle are distinctive; previously considered a subspecies of Wool-Rush, *S. cyperinus*, and often hybridizes with it; *S. pendulus* is similar but lacks these dark bases and the scales of the spikelets have a conspicuous green midrib

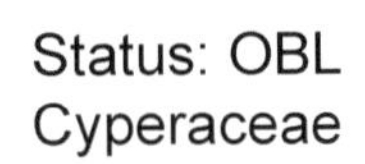

Wool-Rush

Scirpus cyperinuus (L.) Kunth

Plants: grows in dense clumps with fertile upright stems
Leaves: 5-10 per stem; blades 22-80 cm by 3-10 mm; proximal leaves with green to reddish-brown sheaths and few-many, +/- conspicuous septa
Infloresence: terminal; usually 2 x branched; mature branches often spreading or drooping; without auxillary bulblets; bases of invollucral bracts reddish-brown, brownish, or brownish-black
Spikelets: in 2-15 dense cymes; central spikelet sessile, others with or without stalks, 3.5-8.0 mm by 2.5-3.0 mm; scales reddish-brown, brownish, or brownish-black, ovate, 1.1-2.2 mm long, with or without tiny tip
Flowers: 6 perianth bristles persistent, slender, contorted and much longer than achene, projecting beyond scales; mature infloresence appears wooly
Achenes: whitish to pale brown, elliptical or obovate, 3-sided or flattened
Habitat: marshes, moist meadows, ditches, shallow ponds, disturbed areas
Similar Species: lack of truly black bases to involucral bracts distinguish this from Blackgridle Bulrush, *S. atrocinctus*, even before mature spikelets appear; in *S. pedicellatus* the lateral spikelets in a cyme are mostly stalked

Swaying Bulrush

Status: OBL

Shoenoplectus subterminalis

Cyperaceae

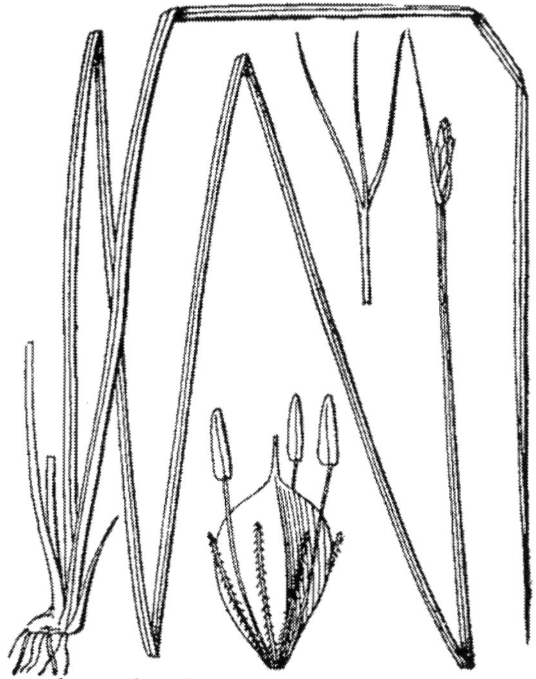

Plants: limp-leaved, mostly submerged plants; flexible stem holds infloresence just above water surface
Leaves: numerous thin limp leaves, usually trailing just under the water surface, to 1 m; occasionally plants may be +/- terrestrial with stiffer leaves
Infloresence: single spike, 7-12 mm long; single bract 1.5-6.0 cm
Spikelets: light brown, 7-15 mm long by 3-7 mm wide; thin scales
Flower: brown, bristlelike
Achene: three angled
Habitat: shallow quiet or flowing waters, marshes
Similar Species: growth form of submerged leaves with only infloresence above water distinguishes this species from other *Scirpus*
aka: *Scirpus subterminalis*; see also Bolboschoenus fluviatilis in the introduction to the Cyperaceae

Softstem Bulrush

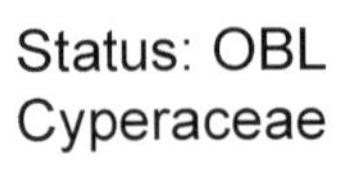

Shoenoplectus tabernaemontani
(K.C. Gmel.) Palla

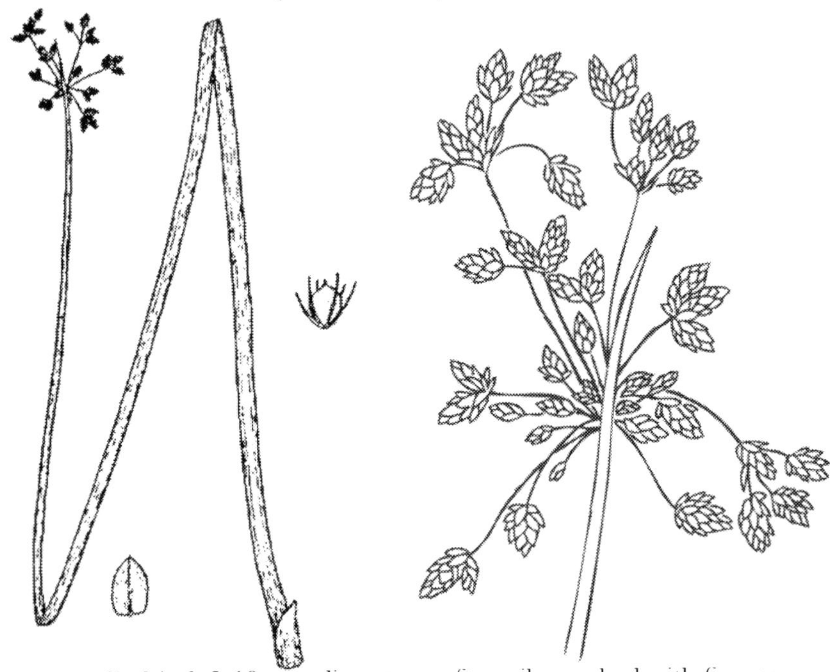

Stems: cylindrical, 3-10 mm diameter; soft, easily crushed with fingers
Leaves: 3-4 basal sheaths, 1-2 have C-shaped to flattened blades, usually much shorter than sheath
Infloresence: 2-4 x branched, branches to 15 cm; proximal bract erect, thickly C-shaped to cylindrical, 1-8 cm long
Spikelets: 15-200, usually solitary or sometimes in clusters of 2-4, reddish-brown; scales uniform dark to pale orange-brown, midrib often pale, margins ciliate, apex round to notched with awn ~ 0.5 mm
Flowers: 6 perianth bristles with retrorse spines about as long as achene
Achenes: plano-convex, dark gray-brown, 1.5-2.8 mm by 1.2-1.7 mm
Similar Species: Part of a complex of hybridizing species that includes Hardstem Bulrush, *S. acutus*, which has a harder, narrower stem (difficult to crush with fingers), spikelets always in clusters and Slender Bulrush, *S. heterochaetus*, is very similar to Hardstem Bulrush (see next page)
aka: *Scirpus validus*

Hardstem Bulrush

Status: OBL
Cyperaceae

Shoenoplectus acutus
(Muhl. ex Bigelow) A. & D. Love

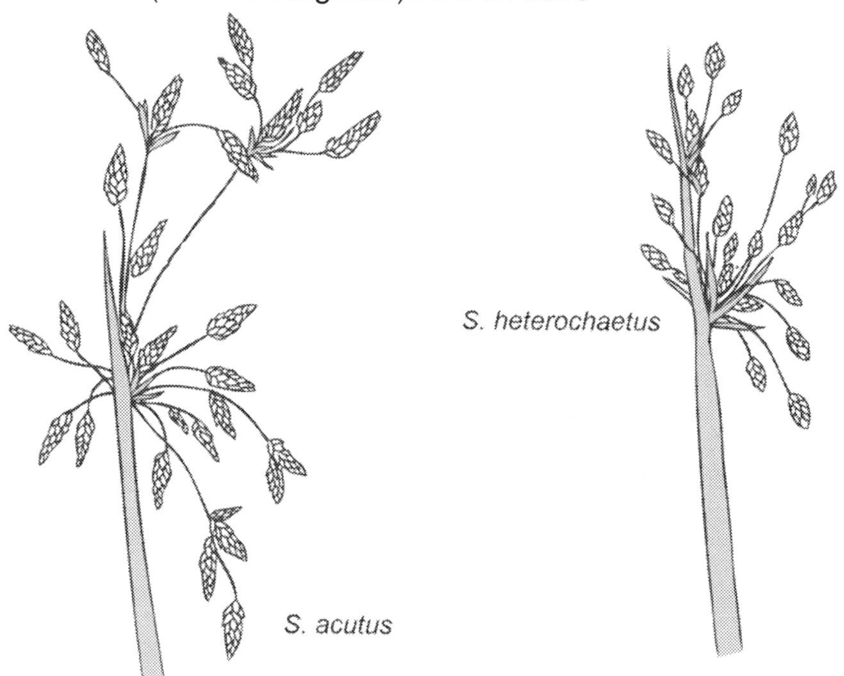

Stems: General form very similar to Softstem Bulrush, S. tabernaemontani; cylindrical, 2-10 mm diameter; firm, not easily crushed with fingers
Leaves: 3-4 basal sheaths, 1-2 have C-shaped to flattened blades, usually much shorter than sheath
Infloresence: 2-3 x branched, branches to 15 cm; proximal bract erect, thickly C-shaped to cylindrical, 2-10 cm long
Spikelets: 3-190, sometimes solitary, usually in sessile clusters of 2-8, gray-brown; scales uniform dark to pale orange-brown, midrib often green when young, margins ciliate, apex round to notched with awn ~ 1 mm
Flowers: 6 perianth bristles with retrorse spines about as long as achene
Achenes: plano-convex, dark gray-brown, 2.0-3.0 mm by 1.2-1.7 mm,
Similar Species: Slender Bulrush, *S. heterochaetus,* is similar to Hardstem Bulrush but with infloresence only twice branched, fewer spikelets (5-30), and a three-angled achene; part of a complex of hybridizing species that also includes Softstem Bulrush, *S. tabernaemontani*, which has a softer, thicker stem (easily crushed with fingers)

Chairmaker's Bulrush

Status: OBL
Cyperaceae

Shoenoplectus americanus (Pers.)

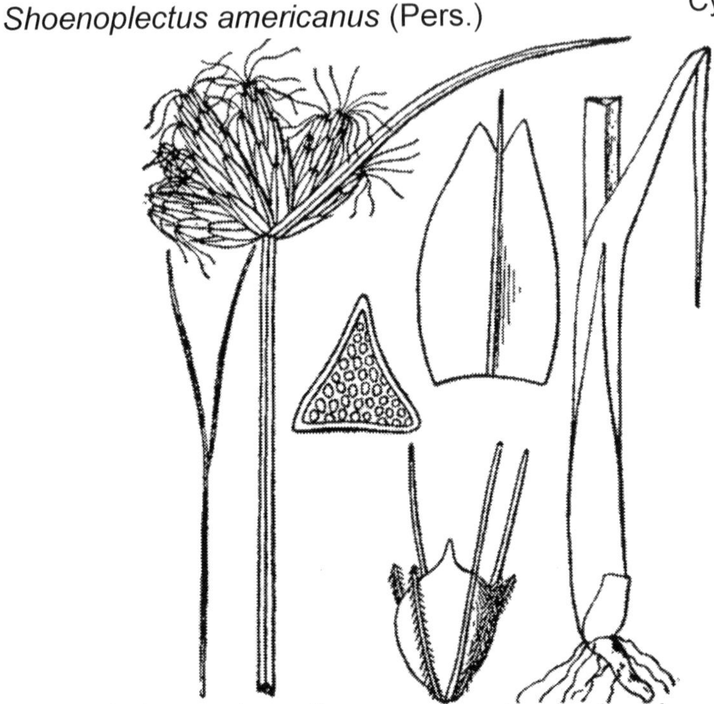

Plants: single stems in small groups; stems sharply triangular, somewhat concave; 0.4- 2.3 m by 2.5-6.0; from rhizomes
Leaves: 3-5 basal, 2-3 stem; V-shaped near base, three-angled distally; 9-30 cm by 2-7 mm; upper leaf blade equal to or shorter than its sheath
Infloresence: sessile clump of 3-35 spikelets; single leaf-like bract 3-20 cm
Spikelets: spikelets bright-orange to purplish-brown, 6-15 mm by 4-5 mm
Flower: bristles yellowish brown, slender to stout, about as long as achene
Achenes: brown, usually biconvex obovoid, 1.9-2.6 mm long, 1.3-1.7 mm
Habitat: freshwater and brackish marshes
Similar Species: trianglar stems and arrangement of spikelets distinguish this species from other Bulrush except Common Threesquare, *S. pungens*, where the blade of the upper leaf on the stem is equal to or longer than the sheath; superficially similar but without triangular stems are: Weakstalk Bulrush, *S. purshianus*, which the infloresence has a somewhat spreading bract 1/3 to 1/30th the stem length; and Smith's Bulrush, *S. smithii*, which has a erect bract 1/5 to 2/3 of the stem length and erect
aka: *Scirpus americanus*

River Bulrush

Bolboschoenus fluviatilis (Torr.) Sojak

Status: OBL
Cyperaceae

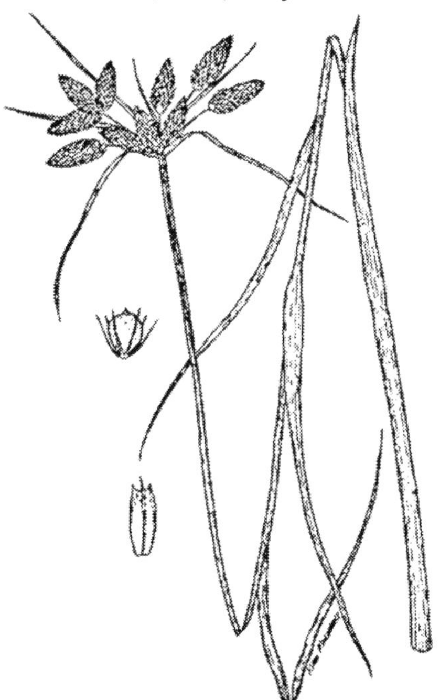

Plants: 60-150 cm tall, erect, triangular stem, from tuber bearing rhizome
Leaves: several elongate linear stem leaves, 6-16 mm wide; sheath convex
Infloresence: terminal; 10-40 spikelets, central sessile, others clumped on lateral branches; 3-5 elongate, drooping leaflike bracts
Spikelets: 12-25 mm long by 6-10 mm wide; with brown scales, 2-4 mm
Flower: yellowish brown bristles
Achenes: three sided, grayish brown, 3.5-5.0 mm
Habitat: freshwater lakes and marshes, tolerates brackish water
Similar Species: drooping bracts, a convex leaf sheath, and the wide spikelets distinguish this species from Scirpus with a similar general appearance
aka: *Scirpus fluviatilis*

Tawny Cottongrass

Eriophorum virginicum L.

Status: OBL
Cyperaceae

Stem: 40-120 cm long by 0.6-1.0 mm wide; colonial on long creeping rhizomes

Leaves: blades flat, ends triangular in cross-section; blade longer than the leaf sheath; to 30 cm long by 1.5-4.0 mm wide

Infloresence: terminal, 2 (3) involucral bracts, leaflike, 4-12 cm long

Spikelets: 2-10, in dense head; ovoid, 6-10 mm flowering, 10-20 mm fruiting; on rough-to-touch 2-10 mm stalks; scales brown, often with green center, with 3-7 nerves, ovate-oblong, 4-5 mm long

Flower: each in the axil of a scale; perianth bristles 10+, tawny-brown, at least at base, 12-18 mm long

Achenes: dark brown to black

Habitat: bogs, wet meadows

Similar Species: this species is distinguished by the tawny-brown bristles of the flower; in the other species: *E. vaginatum*, has a single spike, the others 3 or more; *E. viridicarinatum*, has 2-3 involucral bracts, the remaining have one; in *E. tenellum*, the blade of the uppermost leaf on the stem is at least as long as its sheath and in *E.gracile*, the blade is shorter than its sheath

Needle Spikerush

Status: OBL

Eleocharis acicularis (L.) Roemer & J.A. Schultes Cyperaceae

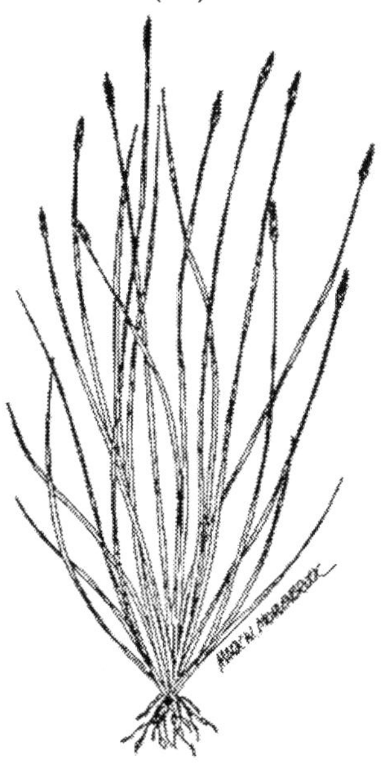

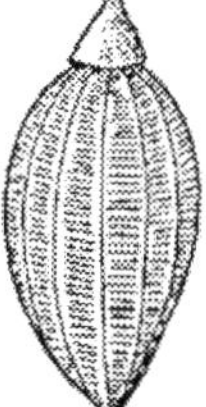

Plants: rhizomes thread-like, 0.25-0.50 mm thick; stems erect to somewhat limp, smooth or with 3-12 ridges, round to somewhat compressed, 1-60 cm by 0.2-0.5 mm; may form partially submerged mats with limp stems
Leaves: distal sheaths may persist or not, straw-colored to red at base, straw-colored to whitish at upper end; end of sheath tight or inflated, often splitting, apex rounded to acute; no blades
Spikes: ovoid to cylindric, 2-8 mm by 1-2 mm, at least twice as thick as stem; 4-25 scales, bright reddish to purplish-brown to straw-colored, often with greenish midrib; 1.5-2.5 mm by 1.0-1.5 mm, apex blunt to acute
Flowers: bristles usually absent; stamens 3, yellowish to brownish; style usually 3 branched
Achenes: about twice as long as wide, 0.7-1.0 mm by 0.35-0.60 mm; three-angled; tubercule constricted at top
Habitat: bare wet soil in meadows, disturbed places or in shallows of ponds, springs, vernal pools
Similar Species: this is the only species with achenes ~ twice as long as wide; the thread-like rhizomes are also distinctive

graminoids: sedge family

Flatstem Spikerush

Eleocharis compressa Sullivant

Status: FACW+
Cyperaceae

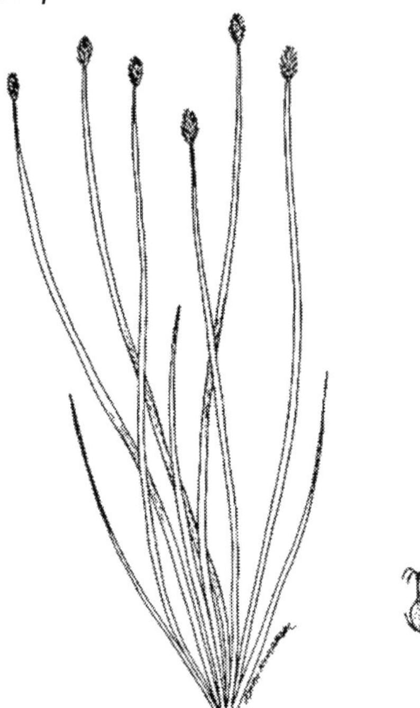

Plants: perennial, mat-forming; rhizomes 2-3 mm thick
Stems: compressed (2-5 x as wide as thick), often with 2-12 ridges; 8-45 cm by 0.5-1.8 mm wide
Leaves: distal sheaths persistent; proximally red, distally green to light-tan
Spikes: 4-8 mm by 2-4 mm; 20-60 scales, ~6 per mm of spikelet; scales spreading in fruit, straw-colored to medium brown to dark brown, midrib often paler; 2-3 mm, apex acute to awl-shaped
Flowers: bristles 0-5, with spines
Achenes: usually fall with scales yellow- to dark-brown, 3-angled; 0.8-1.1 mm by 0.6-0.8 mm; tubercule small
Habitat: damp soils to seasonal shallow waters in calcareous areas
Similar Species: the only species in the area with a compressed stem; two other closely related species have the combination of spike wider than stem/achene 3-angled/achene width ~ 2/3 of length, but with roundish stems: in *E. tenuis* the achenes fall before the scales; in *E. elliptica* the achenes remain after the scales fall; one additional species, *E. intermedia*, is an annual and has no rhizomes

Common Spikerush

Status: OBL

Eleocharis palustris (L.) Roemer & J.A. Shultes

Cyperaceae

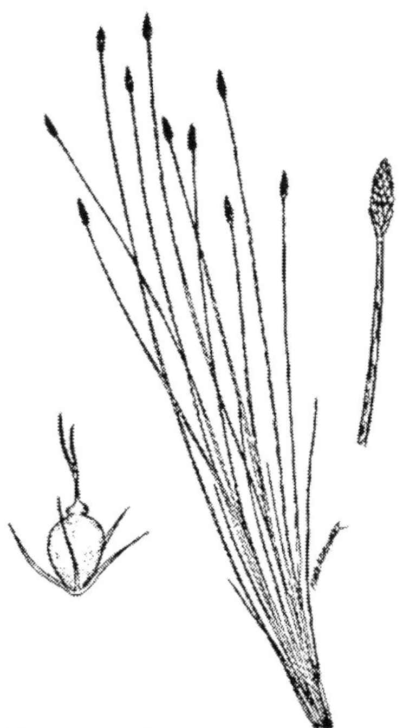

Plants: forms mat with scattered clumps of stems from rhizomes 1.5-4.5 mm thick; stems round to slightly compressed, 30-115 cm by 0.5-5.0 mm
Leaves: distal leaf sheaths persistent or may disintegrate; end abruptly, often split; red to black at base, green or red distally, not inflated
Spikes: 5-25 mm by 3-7 mm; 30-100 scales, 4-8 per mm of rachis, brown with straw-colored to green midrib, 3-5 mm long, acute to subacute, distal ones with keel; the basal scale of wraps ~ 2/3 of the way around the rachis
Flowers: 4-5 bristles (sometimes absent); styles usually 3-branched
Achenes: not persistent on spike, straw-colored to brown, flattened, 1.1-2.0 mm wide by 1.0-1.5 wide; tubercule 0.4-0.7 mm long, constricted at base
Habitat: marshes, wet meadows, shores, ponds
Similar Species: Bald Spikerush, *E. erythropoda*, has a basal scale on the spikelet that wraps ~ 3/4 of the way around the rachis and Yellow Spikerush, *E. flavescens* has a leaf sheath that is prolonged into a tattered whitish tip; the next two do not grow in rhizomatous mats: Ovate Spikerush, *E. ovata* has a tubercule 0.3-0.5 mm wide and Blunt Spikerush, *E. obtusa*, has a tubercule 0.5-0.8 mm wide

Robbin's Spikerush

Eleocharis robbinsii Oakes

Status: OBL
Cyperaceae

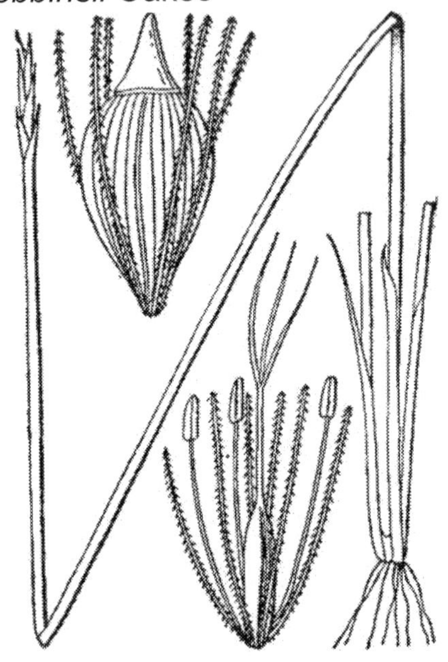

Plants: dense tufts or mats from rhizomes 0.7-0.9 mm thick; emergent stems sharply 3-angled, 16-70 cm by 1-2 mm; when submerged stems form mats of limp sterile stems 0.1-0.3 mm thick
Leaves: upper leaf sheaths persistent or not; no blades
Spikes: 9-33 mm long by 1.5-3.0 mm wide; 4-8 flowers; scales few, lance-shaped, membranous toward margin, apex narrowly rounded to acute
Flower: 6-7 toothed bristles, reddish-brown, longer than achene
Achenes: flattened, straw-colored to medium brown, 2.0-2.5 mm; tubercule ~ one-half length of achene
Habitat: mud or shallow waters of lakes and ponds with sandy to peaty soils
Similar Species: this is the only species in the Adirondacks with the spike only ~ as wide as the stem

graminoids: sedge family

White Beaksedge

Status: OBL

Rhynchospora alba (L.) Vahl

Cyperaceae

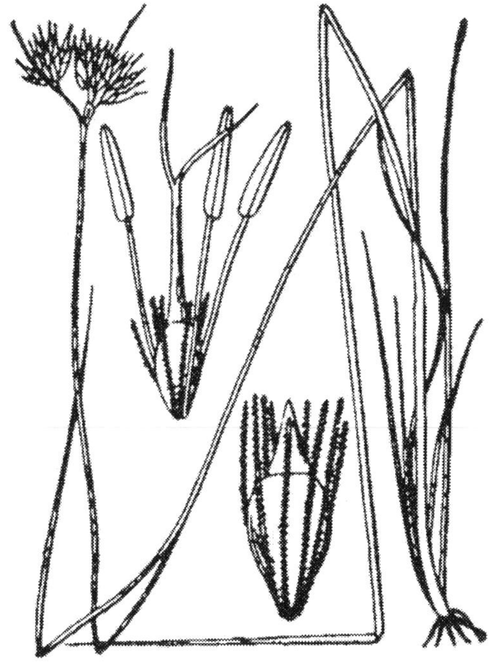

Plants: erect, clustered; stems 6-70 cm, thin, weak, overtopping leaves
Leaves: narrow to hair-like, 0.5-1.5 mm; flat at base, tapering to 3-angled tips
Infloresence: 1-3 irregular clusters of spikes, 1.5-2.5 cm thick; subtending bracts overtopping clusters
Spikes: whitish, 4-5 mm; 2-3 flowers producing 1 achene; scales 4-5 mm
Flower: 8-14 stout bristles, stout, overtopping achene, back pointing barbs, minute bristles at base
Achenes: flattened, 2.5-3.0 mm; tubercule one-half to two-thirds length achene
Habitat: sphagnum bogs and open conifer swamps
Similar Species: the whitish spikes are distinctive, other species are brownish (they also have 5-6 bristles in the flower); Brownish Beaksedge, *R. capitatellata*, is similar but has brown spikelets; Horned Beaksedge, *R. capillacea*, has small (1-10 spikes) clusters, ovoid in shape, and achenes that are less than half as wide as long; Brown Beaksedge, *R. fusca* has no barbs on the bristles

graminoids: sedge family

Smooth Saw-grass Sedge

Cladium mariscoides (Muhl.)

Status: OBL
Cyperaceae

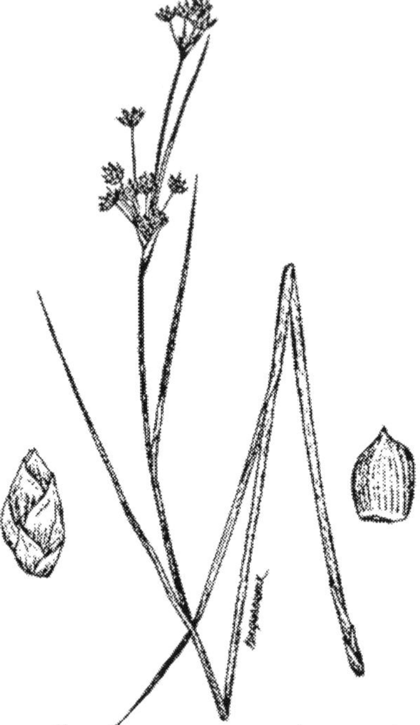

Plants: grows singly, spreading from runners; stems somewhat 3-angled, stiff; 30-100 cm by 1-2 mm

Leaves: few stem leaves; blade V-shaped toward base, rounded distally, smooth margins; 2.0-3.5 mm wide

Infloresence: terminal or terminal and lateral combined; branched cymes, 15-25 cm; forming terminal clusters of spikes

Spike: 3-5 per cluster; lance-shape to narrowly ellipsoid; scales 5-6, 2.5-3.0 mm by 2.0 mm, conspicuous midvein; lowermost scale without flower; middle scales with staminate or sterile flowers; uppermost flower bisexual

Flower: uppermost bisexual; no bristles

Achenes: white to pale green, rounded, 2.5-3.0 mm, conspicuously pointed on top; no tubercule

Habitat: swamps and marshes, peat bogs, wet places

Similar Species: similar to Beak-Rushes, *Rhynchospora*, which have a weaker, thinner stem and prominent tubercules on achenes

graminoids: sedge family

Straw-Colored Flatsedge

Cyperus strigosus L.

Status: FACW
Cyperaceae

Plants: from rhizomes, growing separately or in scattered colonies; stems 20-40 cm tall by 1-6 mm wide, from swollen base, triangular
Leaves: 20-40 cm long by 1-4 mm wide; flat; crowded toward base
Infloresences: terminal; 1-4 branches, each branched (and sometimes rebranched), each forming a cluster of rays (10-40 mm long); each ray bears heads of 12-50 spikes on the upper portion; heads, 10-40 mm tall by 10-28 mm wide; spikes mostly spreading at ~ right angles from the axis; 5-7 leafy bracts, 10-30 cm long, ascending at 30-45 degrees
Spikelets: strongly compressed, 5-30 mm long by 0.6-0.9 mm wide; scales not spreading, green central vein with straw colored sides; 3.2-4.5 mm long, overlapping 1/4-1/2 of next scale; basal scale empty
Flowers: terminal bisexual; 3 stigmas
Achenes: trigonous, 1.8-2.4 mm
Habitat: pond shores, ditches, damp soils, garden weed
Similar Species: seven other species in the Adirondacks: *C. bipartitus, C dentatus, C. diandrus, C. houghtonii, C. lupulinus, C. schweinitzii, and C. squarrosa*

Three-Way Sedge

Dulichium arundinaceum (L.) Britton

Status: OBL
Cyperaceae

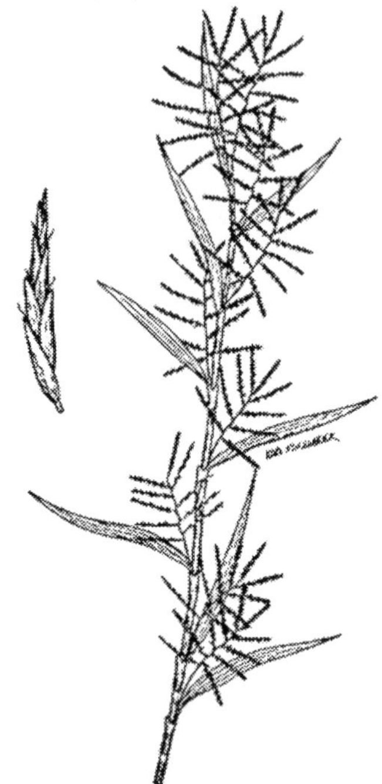

Plant: erect unbranched stems, hollow, 20-100 cm; stem leaves arranged in three distinct rows, leaf attachment either clockwise or counterclockwise on stem
Leaves: basal leaves bladeless; upper leaves (3-15) with short leathery blade, ligule present; blades 4-15 cm x 1-8 mm; upper leaves may serve as primary bracts for infloresence
Infloresence: 6-30 in leaf axils (which act as primary bracts); 3-10 spikes per ray, each 2-5 cm long
Spikes: 1.5-3.0 cm long; 4-8 scales arranged in two ranks
Flowers: bisexual, 6-9 barbed bristles, slightly overtopping the achene
Habitat: open wet places
Similar Species: axillary flattened spikes and infloresences are distinctive

Genus *Carex*

With 119 species found in the Adirondacks, this genus is the most diverse of the family and is also often dominant in wetlands. However, members of this genus are also often the most difficult plants you will have to identify: stalwart field biologists have been know to shudder when they realize they are going to have to identify a sedge.

But they are important. Less is known about the distribution and ecology of these species. This is an area where an enthusiastic amateur can make a real contribution.

In a field guide it is not possible to include every species of *Carex* you may encounter nor to give enough detail that you can reliably identify it. We compromised. The genus is broken down into a number of "superspecies" and we have tried to show you examples of the most common members of these. Note that we have not been able to show all of the superspecies in the Adirondacks. You should try to find a page that most closely matches the specimen you have. Hopefully this will allow you to assign the specimen to the right superspecies; with a close reading of the page you may even be able to assign it to the right species. You should always collect a specimen of the sedge so you can identify it in the laboratory. Since not everyone has access to reference books a key to the members of the genus *Carex* found in the Adirondacks has been made available: www.denniskalma.com/keys/keys.html

This page and the next depicts some of the anatomy and terminology you will need to use this or any other key to *Carex*.

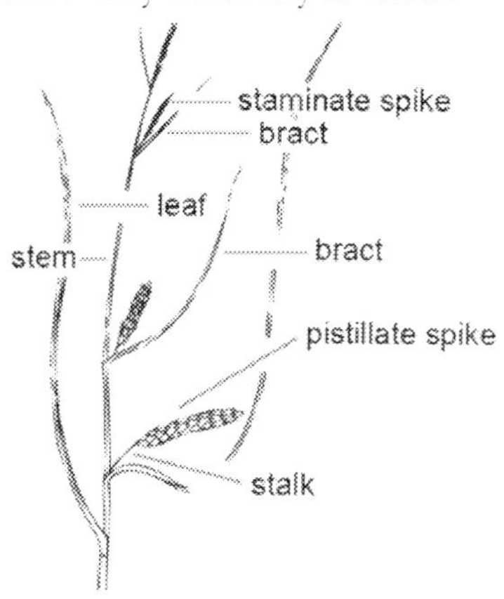

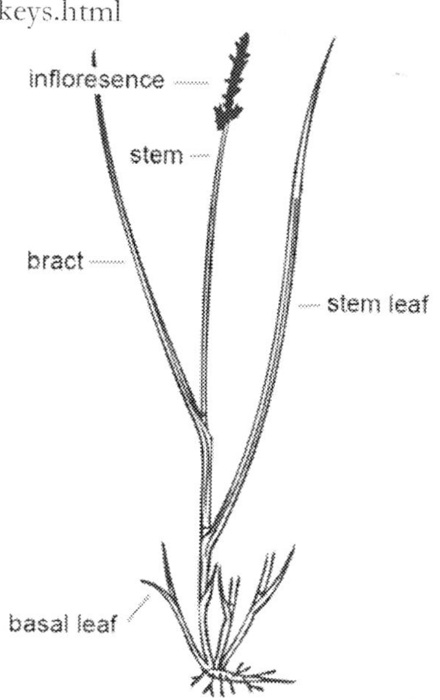

Stem (culm): Stalk that bears the flowering structures of the plant or may be sterile
Leaf blade: The part of the leaf not attached to the stem
Leaf sheath: The part of the leaf wrapped around the stem
Ligule: Collar-like structure around stem where leaf blade joins sheath
Basal leaves: Arise from base, may or may not have blade
Stem leaves: Arise from the stem
Infloresence: The arrangement of the flowers on the stem
Primary bract: Leafy structure coming off the stem just below the lowermost spike of the infloresence
Secondary bract: Bract above the next spike (going up the stem)
Spike: Collection of flowers along a central axis; either staminate or pistillate, or both on separate portions of the axis
Flowers: Either male or female in *Carex*, with two (**bifid**) or three (**trifid**) **stigmas**
Scale: Each flower has a scale coming off the axis just below it; arranged spirally they give the spike a cylindrical shape; the structure of the scales, especially the pistillate scales is often important for identification
Perigynium (plural **perigynia**): Sack-like structure surrounding the achene
Achene: One seeded fruit; style may or may not be hard and persist in the mature achene; may be three-angled or flattened in cross-section

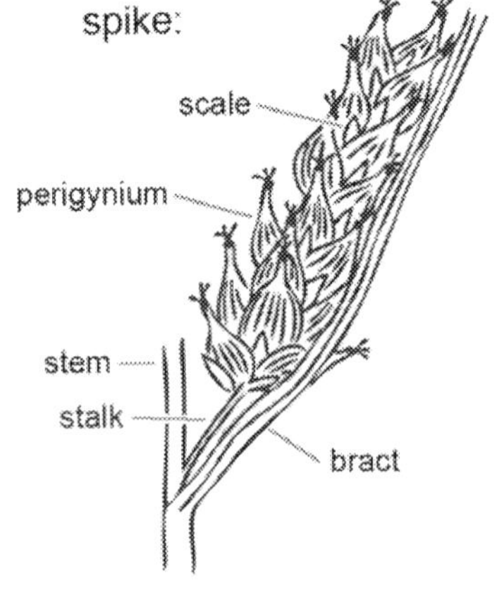

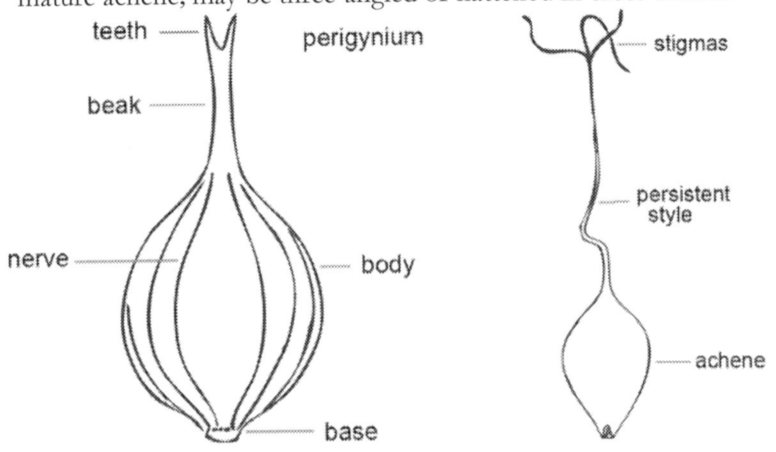

supergenus *Bracteosae* - 7 species graminoids: sedge family

Softleaf Sedge

Carex disperma Dewey

Status: FACW+

Cyperaceae

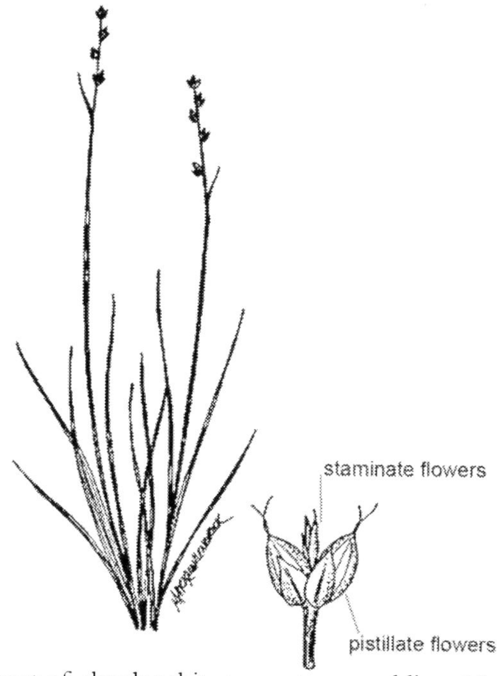

Plant: loose mat of slender rhizomes; stems nodding, 15-60 cm long, exceding leave; previous year's leaves at base
Leaves: basal sheaths pale, truncate at summit; bladed leaves mostly near base; ligules broader than long; blades mid- to dark-green, 15-30 cm by 0.75-1.5 mm, shorter than the flowering stem
Infloresence: infloresence 1.5-2.5 cm long; primary bract 5-20 cm, distal bracts scale-like
Spikes: 2-4, proximal separate, distal aggregated; each 2-5 mm by 2-4 mm; pistillate scales papery-white with green center, ovoid, shorter than perigynia
Perigynia: 2 stigmas; 1-6; plump, greenish to brownish, dotted with whitish transparent pits; 2.25-3.0 mm long by 1.3-1.5 mm wide
Achenes: flattened; glossy red-brown, elliptical, 1.5-1.75 mm by 1.0 mm
Habitat: swamps, bogs, wet meadows, moist coniferous woods
Similar Species: differs from other *Bracteosae* by possessing perigynia covered with whitish dotted pits; other species are *C. appalachia*, *C. cephalophora*, *C. muhlenbergii*, *C. radiata*, *C. retroflexa*, and *C. rosea*

Fox Sedge

Carex vulpinoidea Michx.

Status: OBL
Cyperaceae

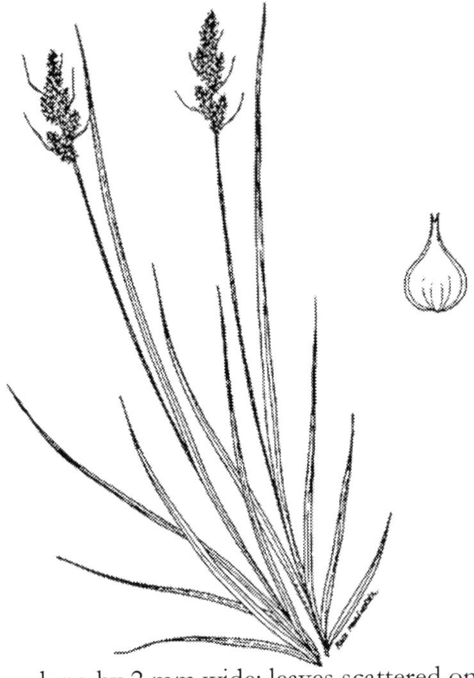

Stem: 30-100 cm long by 2 mm wide; leaves scattered on lower one-half to two-thirds of the stem; grows in loose clumps
Leaves: sheath fronts ridged, spotted pale-brown or red-brown; blades flat, to 120 cm by 5 mm, some overtopping the flowering stem
Infloresence: 7-10 cm long by 15 mm wide; 10-15 branches; compound proximal branches are distinctly separate
Spikes: bracts thin, needle-like, proximal ones conspicuous; pistillate scales brown with awn to 3 mm
Perigynia: 2 stigmas; green to pale-brown; flattened or planoconvex; veinless or sometimes with 3 veins on outside; 2.0-3.2 mm long by1.3-1.8 mm wide; beak 1/3 to 1/2 length of perigynium
Achenes: 2 styles, achene red-brown, ovate, glossy
Habitat: seasonally saturated or inundated soil in open habitats - wet meadows, marshes, roadside ditches
Similar Species: the very similar Yellow-Fruited Sedge, *C. annectens,* has a flowering stem overtopping the longest leaves and yellow-orange perigynia; Awl-Fruited Sedge, *C. stipata,* has a wider stem (~7 mm) and a flowering stem longer than the leaves (see next page)

Awl Fruited Sedge

Carex stipata Muhl. ex. Willd.

Status: OBL
Cyperaceae

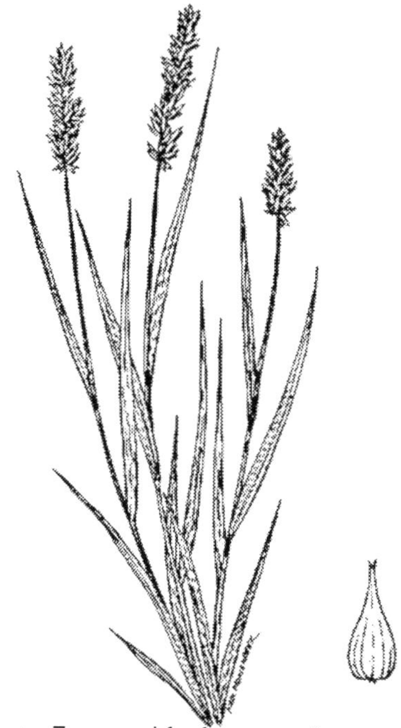

Stem: to 120 cm long, to 7 mm wide; base can be compressed with fingers; basal sheaths of leaves from previous year not persistent; blades to 60 cm by 10 mm, not overtopping infloresence, fronts of blades wrinkled
Infloresence: elongate, to 10 cm
Spikes: 15-25 spikes crowded closely together; scales thin and translucent, without awns
Perigynia: 2 stigmas; 4-5 mm long, 2 mm wide, with spongy base; beak to 2.5 mm; pale brown with ~15 reddish-brown veins on side away from central axis, ~7 on the side toward the central axis
Achenes: flattened, oval-shaped; style persistent
Habitat: seasonally saturated or inundated soils, wet meadows, marshes, edges of swamps, alluvial bottomlands
Similar Species: distinguished from other Vulpinae by combination of two characters: spongy base of perigynia plus sheath prolonged beyond base of blade; Fox Sedge, *C. vulpinoidea,* (previous page) has awns on the ends of the pistillate scales; its leaves are narrower and usually longer than the flowering stalks

Lesser Panicled Sedge

Status: OBL
Cyperaceae

Carex diandra Schrank.

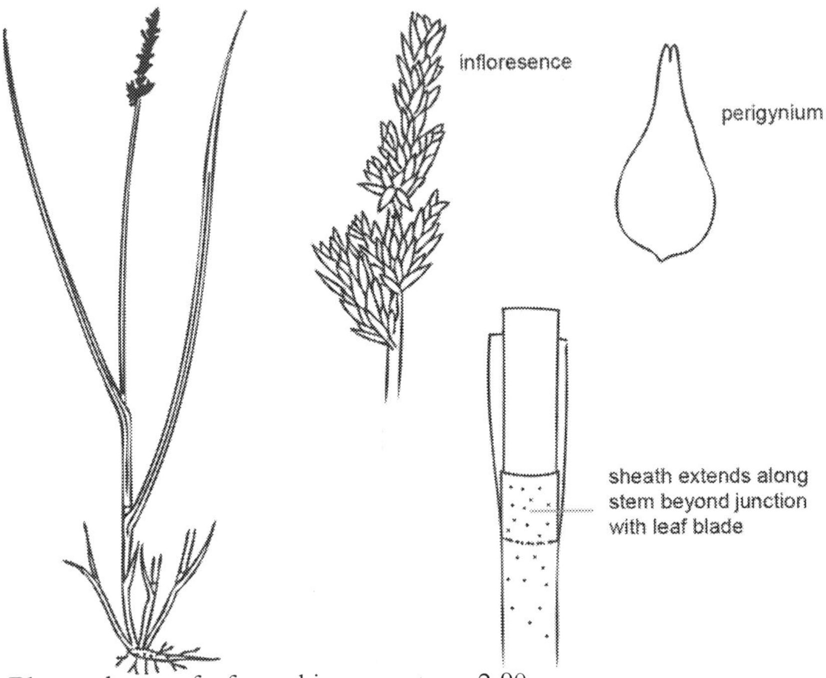

Plants: dense tufts from rhizomes; stems 2-90 cm
Leaves: inside of leaf sheaths (side toward the stem) whitish with reddish dots; sheath prolonged 0.4-4.0 mm beyond the base of the blade; blade 14-30 cm by 1.0-2.5 mm
Infloresence: 2-5 cm; at least one basal branch rebranched; each branch with 2-12 spikes
Spikes: terminal spike staminate; lateral spikes either pistillate or with staminate flowers above pistillate flowers; pistillate scales 1.5-2.7 mm, as wide as perigynia at base
Perigynia: 2 stigmas; spreading; shiny brown, thick walled; flattened; 2.3-2.5 mm by 1.0-1.4 mm, beak 0.9-1.1 mm; 4-6 veins on side toward axis
Achenes: broad, flattened; 1.4-1.7 mm by 0.7-1.0 mm
Habitat: wet meadows, fens, shores and shallow water of peaty lakes
Similar Species: in this superspecies Prairie Sedge, *C. prairea*, has copper-colored inner leaf sheaths; distinguished from supergenera *Multiflorae* and *Vulpinae* by having a perigynium with a thick wall and a more abrupt beak

supergenus *Heleonastes* - 5 species

graminoids: sedge family

Threeseeded Sedge

Status: OBL

Carex trisperma Dewey.

Cyperaceae

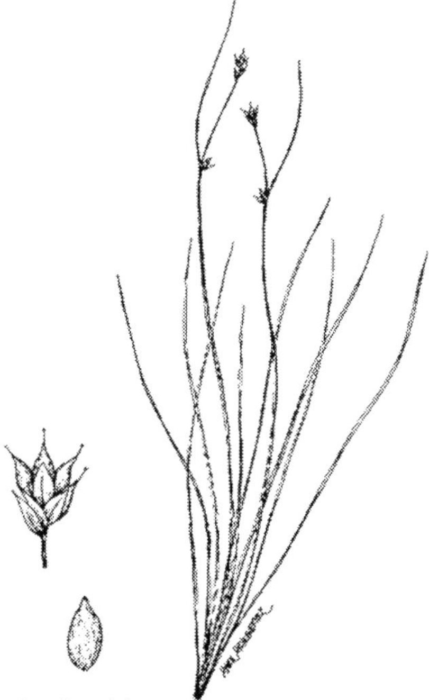

Plants: loose clumps, slender rhizomes; stems weak, arching; 20-70 cm
Leaves: sheaths pale brown on outer-side, thin papery on inner; ligules as long as wide; blades flat or channeled, 5-20 cm by 1-2 mm
Infloresence: 5-10 cm tall; bracts bristle-like, 3-10 cm, overtopping infloresence
Spikes: (1) 2 (3), separated by 2-4 cm, each with 2-5 perigynia; pistillate scales whitish-green, just shorter than perigynia
Perigynia: 2 stigmas; ovate-elliptical, 2.5-3.7 mm by 1.5-2.0 mm; beak entire or with a few small teeth; pale green to brownish with many small veins
Achenes: glossy brown, flattened 1.75-2.0 mm x 1.25 mm
Habitat: sphagum bogs, wet woods, lowlands
Similar Species: this species now separated from Billing's Sedge, *C. billingsii,* which has narrower (0.3-0.5 mm), V-shaped, leaf blades and only 1-2 perigynia in each spike; other *Heleonastes* have either closely aggregated spikes or 4-9 spikes each with 5-30 perigynia: Sparse-Flowered Sedge, *C. tennuiflora*; Brome-Like Sedge, *C. brunnescens*; and Hoary Sedge, *C. cannescens*

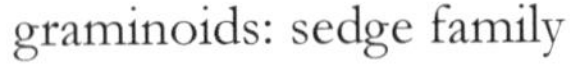

Prickly Bog Sedge

Carex atlantica L.H. Bailey

Status: FACW+
Cyperaceae

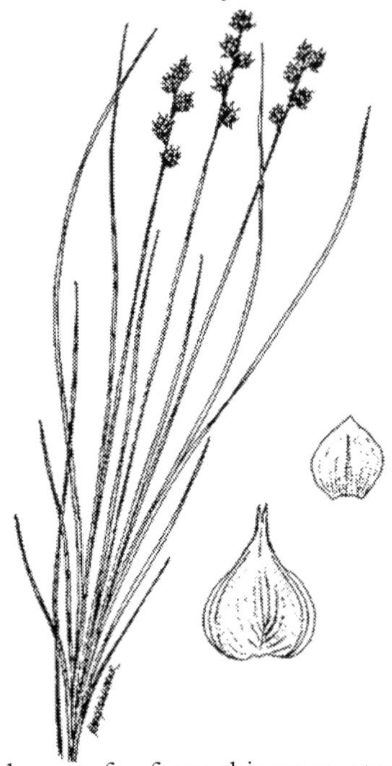

Plant: in low dense tufts, from rhizomes; stem brown at base, 10-110 cm, overtopping leaves

Leaves: sheaths fitting tightly around stems; blades folded like fan, 8-60 cm by 0.4-4.0 mm

Infloresence: racemose with 2-8 spikes, 0.8-5.5 cm

Spikes: terminal spike 4.8-23.7 mm; staminate at base with 2-21 flowers, 1.8-17.5 mm, with 4-38 pistillate flowers above, 3.0-11.0 mm; lateral spikes 3.1-13.1 mm, with ~10 staminate flowers at base tp 6.5 mm and 3-40 pistillate flowers above, 2.5-12.0 mm; basal 2 spikes separated by 1.5-18.0 mm; pistillate scales ovate to almost round, 1.2-2.4 mm by 1.2-2.0 mm

Perigynia: stigmas 2; spreading to reflexed; 6-13 veins on outside, 1-12 on inside; 1.9-3.8 mm by 2.1-3.0 mm, beak 0.45-1.25 mm, teeth 0.15-0.45 mm

Achenes: flattened spoon-shape; 1.0-1.9 mm by 1.0-1.7 mm

Habitat: bogs, acid swamps, wet pinelands

Similar Species: other *Stellulatae* in Adirondacks are *C. arctata, C. echinata, C. exilis, C. interior, C. serosa, C. sterilis,* and *C wiegandii*

Necklace Sedge

Carex projecta Mack.

Status: FACW
Cyperaceae

Plants: grows in dense clumps; only some stems fertile, 50-90 cm; vegetative stems with leaves evenly spaced along distal one-half
Leaves: 3-6 per fertile stem; 18-40 cm long, 3-7 mm wide
Infloresence: usually bent at first internode, brown; 2.5-6.0 cm; first internode 5-12 mm, second internode 3-11 mm; proximal bracts scale-like, sometimes bristle tipped to 3 cm
Spikes: 8-15 spikes, proximal spikes well separated, globose, base tapered & apex blunt, 9-12 mm long; pistillate scales shorter and narrower than perigynia with green or brown midstripe not reaching to tip
Perigynia: 25-30 per spike; 2 stigmas; ascending to reflexed, flattened, 2.9-4.3 mm long, 1.2-1.6 mm wide, 0.3 mm thick; margin winged, 0.2 mm wide
Achenes: ovate to oblong, flattened, 1.1-1.7 mm long
Habitat: wet areas in woods, grasslands, meadows, thickets, shores
Similar Species: Necklace Sedge's infloresence characteristically bent at first node; also in *Ovales*: *C. alata*, *C. argyrantha*, *C. bebbii*, *C. brevior*, *C. cristatella*, *C. foena*, *C. merritt-fernaldii*, *C. scoparia*, *C. sychnocephala*, and *C. tribuloides*

supergenus *Gracillimae* - 1 species

graminoids: sedge family

Drooping Sedge

Status: OBL

Carex prasina Wahlenb.

Cyperaceae

Plants: in dense clumps; fertile stems 30-80 cm by 0.8-1.1 mm, overtopping leaves
Leaves: 2-3 basal sheaths, bladeless or nearly so; uppper leaves 2-5 mm wide; green on back, papery white on front
Infloresence: lateral spikes stalked, shorter than spike itself, up to 4 cm; terminal spike short-stalked, less than 1 cm; proximal bracts leafy, 2-3 mm wide, longer than spike, sheaths less than 3 mm long
Spikes: 2-4 lateral, 1 per node; each overlapping the one above, drooping at maturity; pistillate, 25-50 perigynia, more densely flowered distally; scales whitish-papery with green midrib, shorter than perigynium, often with long awn; terminal staminate or with a few pistillate flowers at the upper end
Perigynia: 3 stigmas; green to golden-green; 2-ribbed, lance-ovoid tapering to flattened, often bent, beak; 2.5-4.0 mm by 1.0-1.5 mm
Achenes: 3-angled, often with concave sides; 1.3-2.0 mm by 1.0-1.2 mm
Habitat: along streams or in low moist forests
Similar Species: see Fringed Sedge, *C. crinita*, which has 2 stigmas *(page 149)*

supergenus *Extensae* - 3 species | graminoids: sedge family

Yellow Sedge

Carex flava L.

Status: OBL
Cyperaceae

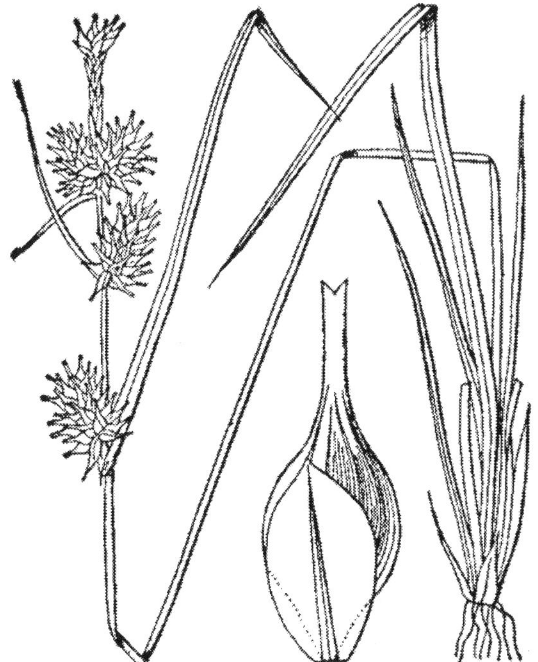

Plants: grows in clumps; stems upright, brown at base, 65-110 cm
Leaves: sheat fronts membranous; blades shorter than stem, flat, to 28 cm by 1.8-3.8 mm
Infloresence: racemose, 2-6 spikes; terminal spike on short stalk 0.4-5.0 mm, lateral spikes sessile
Spikes: 2-5 lateral spikes pistillate; crowded nearer distal portion of spike; 30-40 perigynia per spike, globose-elliptical, 8-22 mm by 2.5-12.7 mm; terminal spike pistillate or rarely with a few perigynia above, 9-22 mm by 1.1-3.0 mm; pistillate scales coppery reddish-brown, 2.3-3.9 mm by 0.9-1.4 mm
Perigynia: 3 stigmas; ascending to reflexed near base; bright yellow when mature; 4.0-6.3 mm by 1.0-1.9 mm, beak 1.3-2.7 mm bent at a 25-70 degree angle
Achene: 3-angled; 1.3-1.7 mm by 0.9-1.2 mm
Habitat: moist to wet habitats, meadows, fens, open swamps
Similar Species: coppery brown scales against yellow perigynia is distinctive; also in supergenus *Extensae*: *C. viridula* and *C. cryptolepis*

supergenus *Hirtae* - 3 species

graminoids: sedge family

Woolyfruit Sedge

Carex lasiocarpa Ehrh.

Status: OBL
Cyperaceae

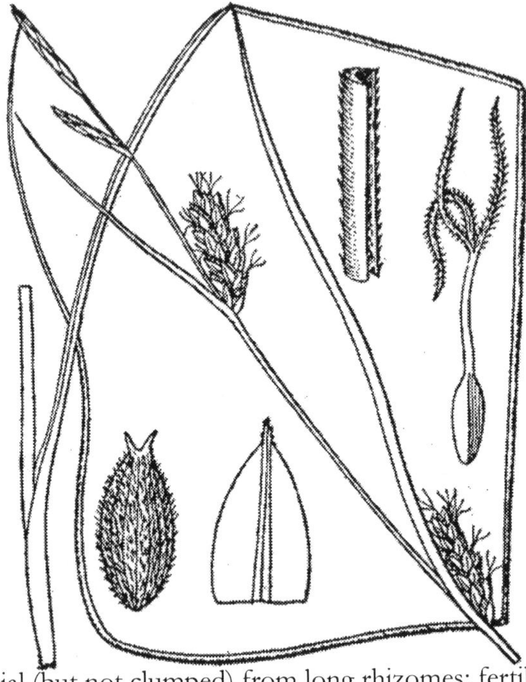

Plants: colonial (but not clumped) from long rhizomes; fertile shoots with stems 40-120 cm, 3-angled, smooth

Leaves: basal sheaths reddish-purple, usually fibrous, bladeless; upper leaves with ligule 1.0-2.3 mm; blades 2.4-4.5 mm wide; gray-green, proximal portions U-shaped, distal becoming V-shaped; blades vegetative shoots ending in a prolonged filiform tip; midvein of leaves (and proximal bracts) form rounded keel

Infloresence: 6-20 cm; 1-2 proximal spikes pistillate, ascending to (more distal) erect; 1-3 terminal spikes staminate, on 2-9 cm stalk; pistillate scales lance- to egg-shaped, acute or sharp-awned, with hairless to rough margin

Perigynia: 3 stigmas; densely pubescent; broad egg-shape, 3.0-4.3 mm by 1.5-2.2 mm, beak short (0.5-1.1 mm); bidentate, straight teeth 0.2-0.7 mm,

Habitat: usually in very wet sites, sometimes forms floating mats, sedge meadows, fens, bogs, lakeshores, streambanks

Similar Species: *C. pellita* has densely pubescent perigynia, but the leaves are wider and M-shaped rather than V-shaped; *C. houghtonia* has longer (4-7 mm) perigynia, conspicuously ribbed, sparsely covered with short hairs

supergenus *Cryptocarpae* - 3 species graminoids: sedge family

Fringed Sedge

Status: OBL

Carex crinita Lam.

Cyperaceae

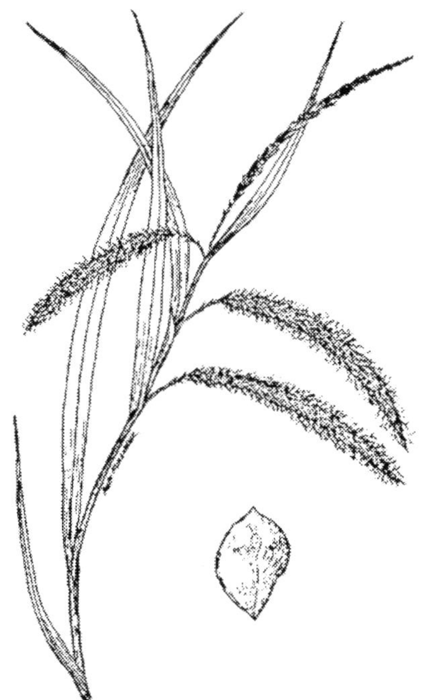

Plants: grows in dense clumps; stem 40-150 cm; 3-angled

Leaves: basal sheaths reddish-brown, without blades; upper leaf sheath hairless, front red-brown, slightly ladder-fibrillose, blade 14-50 cm long, 3.3-10.3 mm wide

Infloresence: raceme; 1-3 proximal spikes pistillate, stalked, pendant; terminal spike (and sometimes distal portion of penultimate) staminate; proximal bract longer than infloresence, 3.7-10.9 mm wide

Spikes: pistillate scales (including awn) 3.4-11.2 mm, copper-brown, midvein reaching to apex; broad, square to notched at end; long, rough awn

Perigynia: divergent; 1.8-3.7 mm; pale brown; egg-shaped with broadest end toward top, round in cross-section; whiteish sheen; apex rounded to very short beak

Achenes: 2 stigmas; achene flattened, constricted on one or both margins

Habitat: wet areas: swamps, meadows, marshes, edges of streams & ponds

Similar Species: Mountain Fringed Sedge, *C. gynandra,* has slightly roughened sheaths with minute rough teeth/hairs (the sheath will feel rough when you touch it with the tip of your tongue - in *C. crinita* smooth)

Tussock Sedge

Status: FACW
Cyperaceae

Carex torta Boott ex Tuck.

Stem: grows in clumps; stems 25-75 cm long, angled; arising lateral to last year's growth
Leaves: basal sheaths red-brown; proximal leaf sheaths hairless, unspotted; 3-5 mm wide, "M" shaped in cross-section
Infloresence: upper spikes erect, proximal ones pendant; proximal bract somewhat shorter than infloresence, 1-3 mm wide
Spikes: proximal 3-4 spikes pistillate, usually sessile, 2.5-9.0 cm long by 4-5 mm wide; terminal 1-2 spikes staminate; pistillate scales purple-brown to black with straw-colored central stripe, shorter and narrower than perigynia, acute, awnless
Perigynia: divergent; green, veinless, somewhat flattened, ovoid; 2.3-4.7 mm by 1.1-1.8 mm; beak often triangular and twisted, 0.1-0.3 mm long, orifice oblique
Achenes: 2 styles; flattened
Habitat: clumps in rocky streambeds/streambanks, sandbars
Similar Species: growth habit in streambeds similar to Tussock Sedge, *C. stricta*, but lower pistillate spikes don't droop in that species *(page 152)*

supergenus *Acutae* - 6 species

graminoids: sedge family

Water Sedge

Carex aquatilis Wahlenb.

Status: OBL

Cyperaceae

Stems: plants not in bunches, stems 20-120 cm
Leaves: basal sheaths reddish-brown, bladeless; sheaths of proximal leaves hairless, fronts with indistinct spots, U-shaped; blades 2.5-8.0 mm wide, papillose on both surfaces
Infloresence: 2-7 proximal spikes pistillate, erect, on stalk 1-4 cm long; 2-4 terminal spikes staminate; bract of proximal spike longer than infloresence
Spikes: pistillate scales red- to purple-brown with narrow pale midvein, awnless
Perigynia: 2 stigmas; ascending, ellipsoidal, no veins, somewhat flattened; 2.0-3.6 mm by 1.3-2.3 mm, beak pale- to purple-brown, less than 0.2 mm
Achene: flattened, glossy, not constricted
Habitat: marshes, bogs, wet meadows, shallows along lakes; usually in acidic substrate
Similar Species: Shore Sedge, *C. lenticularis* also has a proximal bract overtopping the infloresence, but the perigynium has a few elevated nerves on the surface

Upright Sedge, Tussock Sedge

Carex stricta Lam.

Status: OBL
Cyperaceae

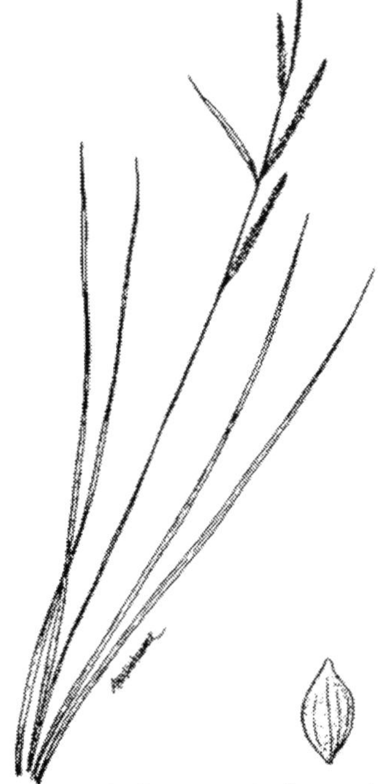

Stem: 50-150 cm, sharply angled; grows in large dense tufts
Leaves: basal sheaths reddish-brown; proximal leaves bladeless; sheaths rough, ladder-fibrous, apex red-brown & U-shaped; blades 4-6 mm, M-shaped
Infloresence: spikes erect; 2-4 proximal pistillate; 2 distal staminate spikes; proximal bract shorter than infloresence, 3.0-4.5 mm wide
Spikes: pistillate 1.6-3.4 cm long; pistillate scales shorter than perigynia, red-brown usually with pale mid-stripe
Perigynia: 2 stigmas; ascending; 1.7-2.4 mm long; pale brown sometimes with red spots on the upper half; somewhat flattened, 0-5 veins on each side; apex acute, beak 0.1-0.2 mm
Achenes: 2 stigmas; achene flattened, not constricted
Habitat: marshes, bogs, wet meadows, shores, especially in seasonally flooded locations, clumps in rocky streambeds
Similar Species: *C. torta* , *C. aquatilis,* and *C. lenticularis* have proximal bracts overtopping infloresence; *C. bigelowii, C. haydenii,* and *C. emoryi* have proximal leaf sheaths that are smooth rather than fibrous

Bottlebrush Sedge

Carex hystericina Muhl. ex Willd.

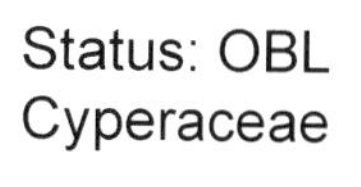

Status: OBL
Cyperaceae

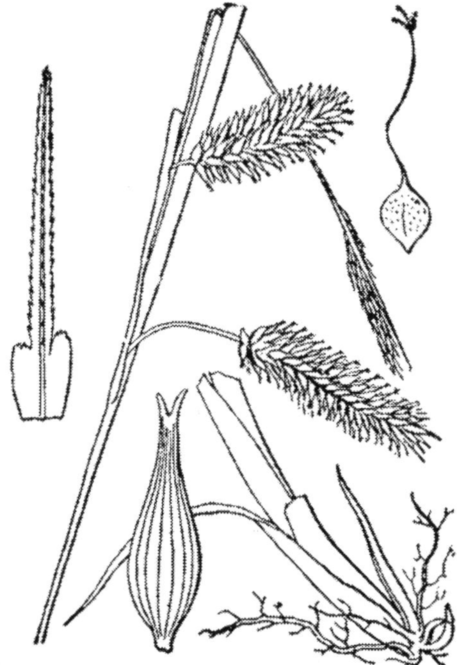

Plants: growth densely to loosely clumped; stems 20-65 cm; 3-angled
Leaves: basal sheaths reddish-purple; ligule longer than wide; leaves 2.5-8.5 mm wide, flat to W-shaped, pale- to mid-green
Infloresence: 2.5-12 cm; 2-3 proximal spikes pistillate, proximal ones spreading to pendant (a characteristic of *Pseudocyperae*); terminal spike staminate; proximal bract 4-30 cm, longer than infloresence
Spikes: pistillate scales 2.3-6.5 mm, narrow-oblong, often cilate margined, apex truncate to notched, with rough awn as long as body; staminate scale rough awned, often cilate margined
Perigynia: 3 stigmas; spreading to somewhat reflexed (proximally); 4.8-6.5 cm, elliptical; 13-21 distinct veins (each separated by 3+ their width), ; beak 1.9-2.8 mm; two straight teeth, 0.3-0.9 mm
Achene: pale brown, 3-angled
Habitat: open swamps, sedge meadows, shores; mostly in calcareous soils
Similar Species: the other two *Pseudocyperae* have perigynia spreading to reflexed on spike: *C. comosa* long and divergent perigynia teeth are long and divergent while *C. pseudocyperus* has short (0.5-1.0 mm), straight teeth

Hairy Sedge

Carex lacustris Willd.

Status: OBL
Cyperaceae

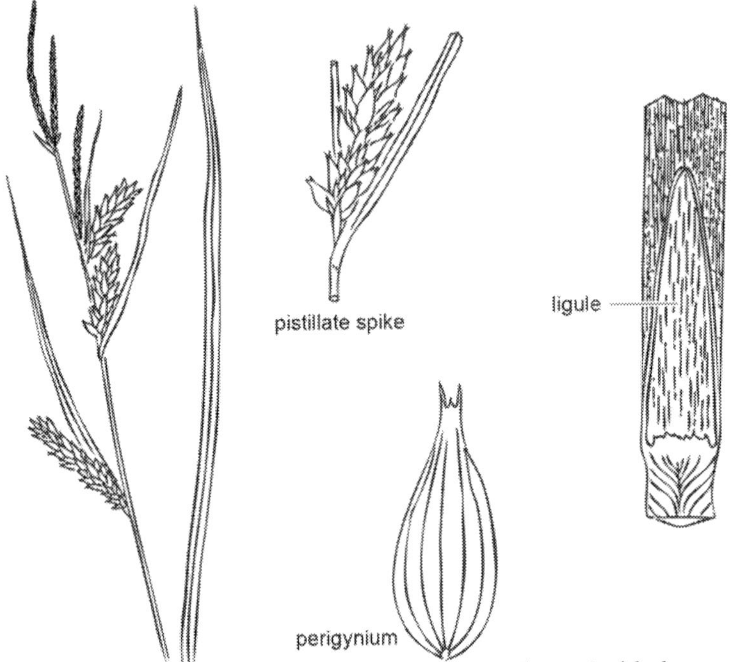

Plants: grows in colonies by rhizomes; stem 50-135 cm; 3-sided, especially near base; often reddish-purple near base

Leaves: basal sheaths fibrous, reddish purple, bladeless; upper leaves with whitish bloom or pale green; ligule 13-40 mm, longer than wide; 8-21 mm wide; M-shaped

Infloresence: raceme, 17-60 cm; proximal 2-4 spikes pistillate, ascending to arching; distal 3-7 spikes staminate; proximal bracts leaf-like, without sheath

Spikes: pistillate scales ovoid with 0.3-3.5 mm awn, more or less rough

Perigynia: 3 stigmas; ascending; ovoid, strongly 14-28 veined, thick-walled, hairless; 5.2-7.8 mm by 1.6-2.5 mm, beak 0.5-1.6 mm, bidentate, straight teeth

Achene: 3-angled; style persistent, straight

Habitat: open swamps, open wet thickets, marshes, shores of lakes, ponds, and slow streams

Similar Species: the thick-walled perigynia place it in the *Paludosae*, distinct from the *Vesicariae* and *Lupinulae*; *C. trichocarpa* has a bigger perigynium, 7-10 mm with longer teeth, 1.5-2.5 mm

supergenus *Vesicariae* - 8 species

graminoids: sedge family

Shallow Sedge

Carex lurida Waldenb.

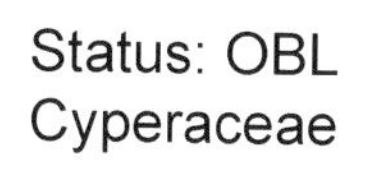

Status: OBL

Cyperaceae

Plants: in loose to dense clumps; stem 25-95 cm, 3-angled, rough distally
Leaves: basal sheaths reddish-purple; blades 4.5-11.5 mm; flat to W-shaped; dark-green
Infloresence: 3-18 cm; proximal bract 9-45 cmm exceeding infloresence; proximal 1-3 spikes pistillate, proximal spreading, distal erect; terminal spike staminate; proximal bract without sheath, 9-45 cm, exceeding infloresence
Spikes: pistillate 15-22 mm thick; pistillate scales narrow, oblong, shorter than the perigynia, apex truncate with a long rough awn; staminate scales with long rough awns
Perigynia: 3 stigmas spreading when mature; 6.5-10.8 mm; 7-12 veins, separate nearly to apex of beak; beak 2.5-5.9 mm, shorter than body; bidentate, teeth 0.2-0.8 mm, straight
Achene: 3-angled, yellow to brown, smooth; style persistent
Habitat: swamps, wet meadows, marshes, wet forests, aquatic margins
Similar Species: of other *Vesicariae* with rough awn on pistillate scales: *C. baileyi* has perigynium beak longer than body; *C. schweintzii* staminate scales without awns

Knotsheath Sedge

Carex retrorsa Schwein.

Status: FACW+
Cyperaceae

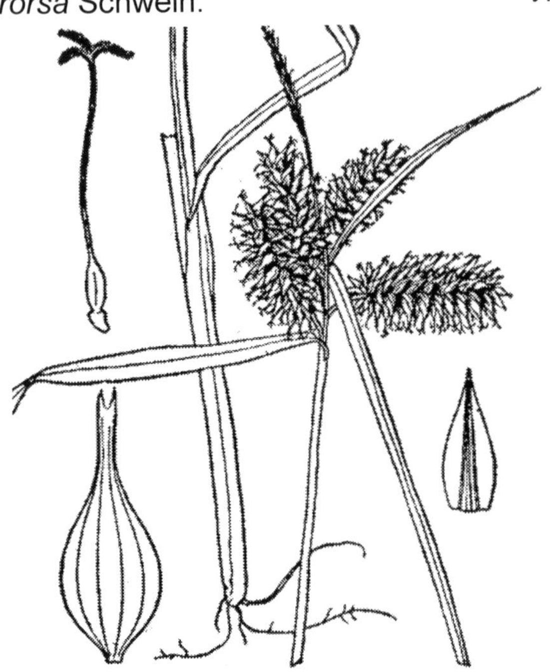

Plants: grows in clumps, rhizomes short to lacking: stem 10-105 cm, 3-angled, smooth distally
Leaves: basal sheaths reddish-brown; widest blades 3-10 mm, flat to W-shaped, mid- to dark-green
Infloresence: 3-20 cm; ; proximal 3-6 spikes pistillate, mostly spreading at maturity but with at least the distal one erect, cylindrical; terminal spike staminate, more or less even with the top of the pistillate spikes; proximal bract without sheath, 3-9 x length of infloresence
Spikes: pistillate spikes with 20-150 perigynia, pistillate scales narrow, oval, ends acute to acuminate
Perigynia: 3 stigmas; mostly reflexed at maturity (more so than other species); 6-10 mm by 2.1-3.4 mm, strongly 6-13 veined with veins running into the beak; beak 2.1-4.5 mm, abrupt; bidentate, teeth 0.3-1.1 mm, straight
Achene: 3 stigmas; achene 3-sided, brown, symmetrical; style persistent
Habitat: swamps, marshes, sedge meadows, along streams and lakeshores
Similar Species: *C. utriculata* has a bract no more than twice the length of the infloresence and the leaves are yellow-green

supergenus *Vesicariae* - 8 species

graminoids: sedge family

Tuckerman's Sedge

Carex tuckermanii Dewey.

Status: OBL

Cyperaceae

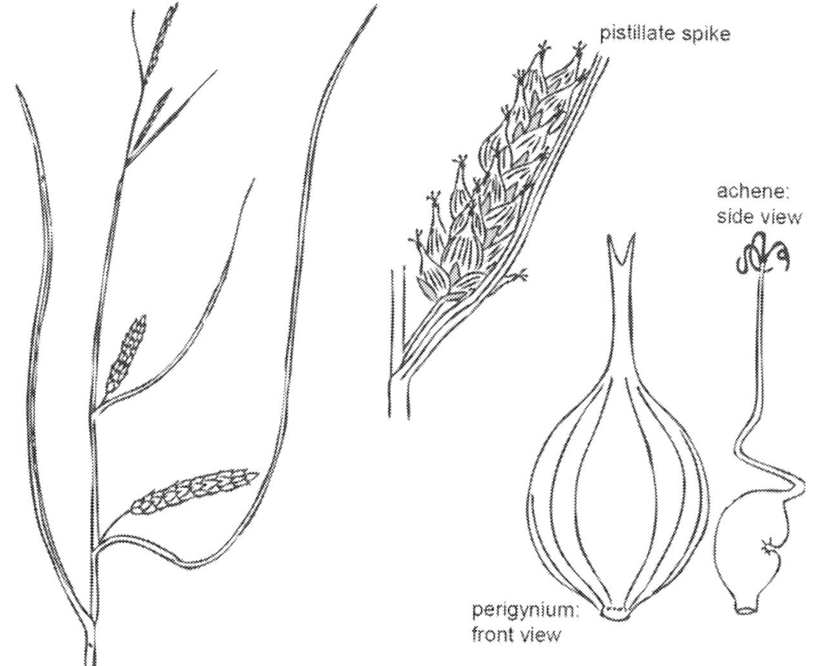

Plants: clumped, from rhizomes, stem 40-120 cm; 3-angled, smooth distally
Leaves: basal sheaths reddish-purple; ligules as long or slightly longer than blade width; blades 2-5 mm; flat to W-shaped; dark-green
Infloresence: 10-35 cm; proximal 2-3 spikes pistillate, proximal spreading to pendulant, distal more erect; terminal spikes staminate; proximal bract without sheath, 25-70 cm, exceeding infloresence
Spikes: pistillate scales narrow, oblong, shorter than the perigynia, apex truncate without awn
Perigynia: 3 stigmas; ascending when mature; ovoid, side view spoon-shaped; 7.5-12.5 mm; green to straw-colored; 7-12 veins, separate nearly to beak; beak 2.4-4.8 mm, smooth, bidentate; teeth 0.7-1.9 mm, straight
Achene: deeply indented on one side, brown, smooth; style persistent, bent
Habitat: deciduous forest swamps, thickets, along streams and ponds
Similar Species: in superspecies *Vesicariae* rather than *Lupulinae* because lack of sheath on lower pistillate bract; large perigynia in tight cylindrical spike and achene indented on one side are distinctive; Few-Seeded Sedge, *C. oligosperma,* has no teeth on beak

Blister Sedge

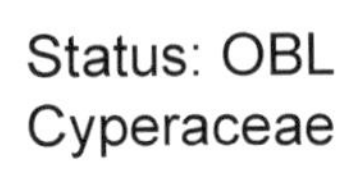
Status: OBL
Cyperaceae

Carex vesicaria L.

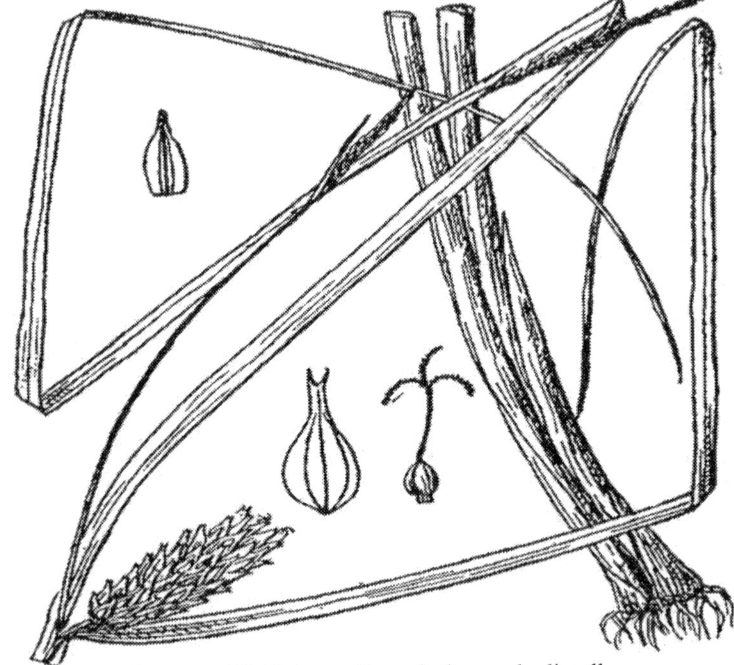

Plants: clumped; stems 15-105 cm; 3-angled, rough distally
Leaves: basal sheaths reddish-purple; ligules longer than wide; blades 1.8-6.5 mm; V- or W-shaped; dark-green
Infloresence: 8-45 cm; proximal 1-3 spikes pistillate, proximal spreading, distal erect; terminal 1-3 spikes staminate, elevated well above summit of pistillate spikes; proximal bract 10-50 cm exceeding infloresence, but not by more than 2.5 x
Spikes: pistillate with 20-150 perigynia; pistillate scales narrow, lance-shaped, shorter than the perigynia, apex truncate without awn;
Perigynia: 3 stigmas; ascending when mature; 4.0-7.5 mm by 1.7-3.5 mm; green to straw-colored; 7-12 veins, separate nearly to apex of beak; beak distinct, 1.1-2.6 mm, smooth; bidentate, teeth 0.3-0.9 mm, straight
Achene: 3-angled, brown, smooth; style persistent
Habitat: swamps, wet thickets, depressions, seasonally innundated areas
Similar Species: in supergenus *Vesicariae* rather than *Lupulinae* because lack of sheath on lower pistillate bract; closest to *C. tuckermanii* but has achene without indentation in side

Great Bladder Sedge

Carex intumescens Rudge.

Status: FACW+
Cyperaceae

Stems: often solitary, sometimes in clumps; stems erect, 30-80 cm
Leaves: 6-12; basal sheaths reddish-purple; sheath of distal leaf 0-1 cm, membranous front; ligule rounded, 1-8mm; blades 8-27 cm by 3.5-8.0 mm
Infloresence: 2-15 cm; stalk of proximal pistillate spike 0.3-1.5 mm; bottom two spikes often so close together that they are difficult to distinguish; stalk terminal spike 0.5-4.0 mm; 1-4 proximal spikes pistillate, terminal spike staminate, 1-5 cm; bracts leafy, sheathless, 6-21 cm by 2-6 mm
Spike: each pistillate spike 1-12 flowered; pistillate scales 1-3 veined, narrow egg-shaped, acute to rough-awned (to 6.5 mm)
Perigynia: 3 stigmas; ascending to spreading, or basal reflexed; 10.0-16.5 mm; strongly 13-23 veined, ovoid (base convex), satiny luster, hairless, beak poorly defined (2.0-4.2 mm)
Achenes: 3-angled, 3.5-5.7 mm; style persistent
Habitat: dry to wet woodlands, forest openings, thickets, wet meadows
Similar Species: *C. grayi*, the other species with very large perigynia, has more flowers in the spikes (8-35), dull perigynia, and style not persistent

Hop Sedge

Carex lupulina (Muhl.) Willd.

Status: OBL
Cyperaceae

Plants: loose clumps or solitary; stems erect, 20-100 cm
Leaves: 4-8; basal sheaths reddish-brown; distal sheaths 1.7-10.0 cm; ligules triangular, 3.5-18.0 mm; blades 15-64 cm by 4-15 mm
Infloresence: 4-40 cm; basal pistillate spikes with 0.5-2.0 cm stalks, spikes 1-20 cm apart; stalk of terminal staminate spike 0.5-6.0 cm; leafy bract with sheath 0.5-15.0 cm and blade 13-55 cm by 3-11 mm
Spikes: 2-5 proximate pistillate spikes, upper more crowded, 8-80 perigynia, ovoid to cylindrical, 1.5-6.5 cm by 1.3-3.0 cm; 1-2 staminate spikes 1.5-8.5 cm by 1-5 mm; pistillate scales lance-ovate, 6-15 mm by 1.0-2.3 mm, rough awn to 6 mm
Perigynia: 3 stigmas; ascending; 13-22 veined, shiny; 11-19 mm by 3-6 mm; beak 6-10 mm
Achenes: 3-angled; rhomboidal with rounded corners, flat to concave faces, 3.0-4.0 mm by 1.7-2.6 mm; style persistent and bent
Habitat: wet deciduous or mixed forest, wet meadows
Similar Species: False Hop Sedge, *C. lupuliformis*, achenes are as long as wide and have sharper corners and edges

Grass Family (Poaceae):

The true grasses are another large and somewhat difficult family of plants found in the Adirondacks. Unlike the sedges they are mainly upland plants. Only a relatively small proportion are found in wetlands. The guide gives illustrations and descriptions for some of the more common species, but you must be aware that it is not a complete list and that you may find that some grasses you encounter will not be covered.

As in the sedge family, the flowers of grasses are highly specialized and, accordingly, have their own terminology. As with most plants much of this terminology deals with the reproductive structures.

The basic unit in the grasses is the **spikelet**: it is a modified flowering branch. It has a central axis called the **rachlla**, which bears (from bottom to top) the **first glume**, the **second glume**, and one or more **florets**.

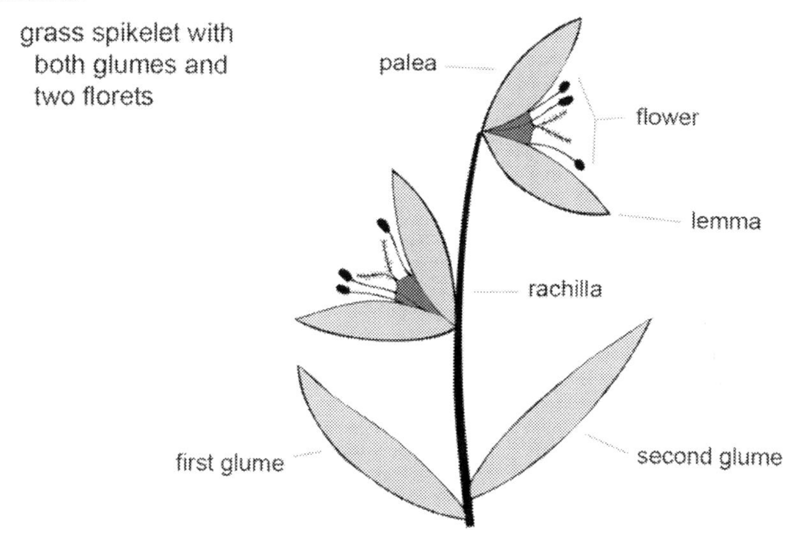

The **floret** consists of the **flower** (the ovary, style, stigma, and three anthers and two spoon-shaped **bracts**. The upper bract, the **palea**, has its back toward the rachilla; the lower, the **lemma**, has its back away from the rachilla. The base of the lemma is called the **callus**; it may be thickened and/or bear hairs or bristles.

Florets can be **bisexual (perfect)**, **staminate (male)**, or **pistillate (female)**. If the sexes are in separate flowers they may be in the same spikelet, on separate spikelets, in different infloresences, or on separate plants.

The **glumes** are empty bracts. They are arranged on alternate sides of the rachilla. Either glume may be reduced in size and the first (the bottom one) may be completely lacking.

The axis of the spikelet runs from the rachilla outward. So if bracts (either the glumes or the palea/lemma) are laterally compressed they are flattened from the sides of this axis. If they are dorsally compressed they are flattened in the direction of this axis.

All of the spikelets are arranged on a usually branched structure called the **panicle**. The panicle and the attached spikelets are called the **infloresence**.

Another set of terms deals with the structures found where the leaf blade arises off from the stem.

The leaf begins at a **node** on the **stem (culm)**. In most grasses the edges of the leaf wrap around the stem for a distance forming a structure called the **sheath**. In the sheath of most of the true grasses the edges of the leaf wrap around the stem and overlap in front - this is an "**open**" sheath (see the genus *Carex* in the sedge family for the definition of a "closed" sheath).

Where the leaf leaves the stem there is often a collar-like structure, the **ligule**, which acts as a seal at the junction. It may be either a membrane or hairs. The portion of the leaf above the ligule is called the **blade**.

Redtop

Agrostis gigantea Roth

Status: FACW
Poaceae

Plants: rhizomatous (not stoloniferous), sod-forming
Stems: 100-150 cm
Leaves: upper ligules 2.5-6.0 mm; blades 3-8 mm wide
Infloresence: panicle 10-20 cm, purplish-red; at maturity conical, widely spreading rough branches, some branches with spikelets at base
Spikelets: crowded, each 2.0-3.5 mm; 1-flowered, articulated above the glumes; glumes about equal, acute, 1-veined, somewhat rough keeled
Florets: lemma acute, ~ 2/3 length of glume; short-awned from the back; minutely bearded; palea 2/3 length of lemma; minute hairs at base
Habitat: moist meadows, shores
Similar Species: Creeping Bentgrass, *A. stolonifera,* is very similar but spreads by stolons and does not form a thick sod; it has a shorter , greenish or straw-colored panicle, usually less than 10 cm tall

Fringed Brome

Bromus ciliatus L.

Status: FACW
Poaceae

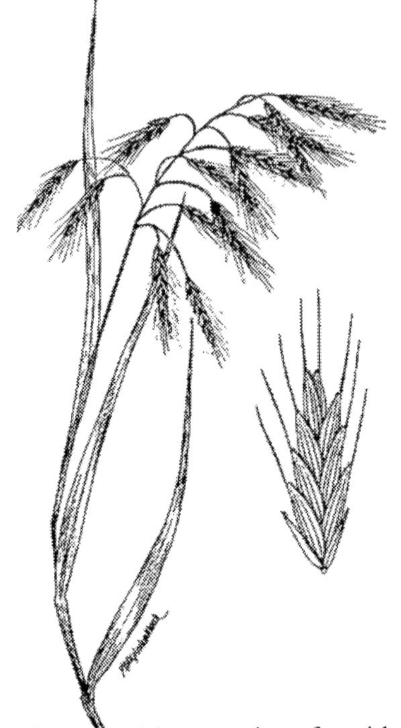

Plants: 60-120 cm tall; perennial; grows in tufts with one-few culms in each tuft
Leaves: 4.0-10 mm wide; flat; elongated, sometimes hairy on the either surface; ligule 0.3-1.0 mm
Infloresence: loose; drooping; 7-20 cm long
Spikelets: drooping; with 4-10 florets, each 15-25 mm
Florets: lemmas 10-13 mm, rounded or only weakly keeled; lemma with awn 3-5 mm; lemma long haired on margins, especially near base
Habitat: fens, swamps, wet meadows, streambanks
Similar Species: Smooth Brome, *B. inermis*, grows from rhizomes, usually in drier habitats and has lemmas hairy all over with only short awns; *B. latiglumis* (aka *B. altissimus*) has a long awn on the lemma but it is hairy all over

Bluejoint

Calamagrostis canadensis (Michx.) Beauv.

Status: FACW+
Poaceae

Plants: perennial, rhizomatous
Stems: 50-150 cm, often branched above
Leaves: sheaths hairless, 3-8 mm; blades rough on both sides, 4-8 mm wide
Infloresence: contracted, later open panicle with spreading branches; 8-25 cm with branches 3-8 cm
Spikelets: 2-6 mm; 1-flowered, articulated above the glumes; glumes about equal or 2nd glume slightly longer, sharp-pointed, 5 inconspicuous veins, tapering to a shredded tip
Florets: lemma smooth, 3/4 to 1 x length 2nd glume, sharp-pointed, 5 inconspicuous veins, tapering to a shredded tip, bearing a slender delicate straight dorsal awn; subtended by numerous white hairs about the length of the lemma that arise from the callus; palea membranous, 1/2 to 1 x length lemma
Habitat: open swamps, wet meadows, moist pockets of soil in mountains
Similar Species: *C. pickeringii* has a more compact infloresence

Sweet Woodreed

Cinna arundinacea L.

Status: FACW+
Poaceae

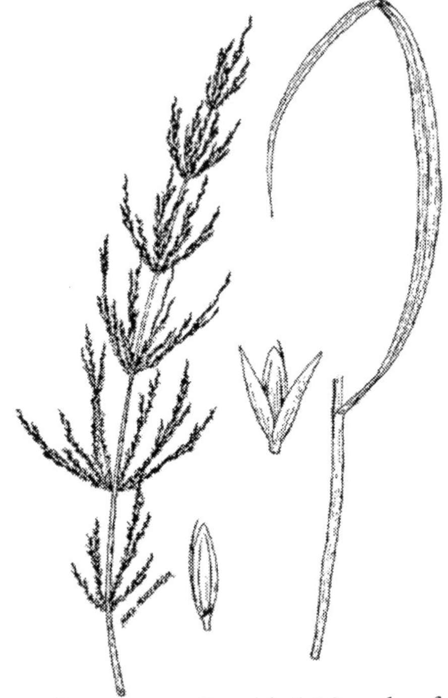

Plants: 100-150 cm tall; erect; usually with 5-10 nodes; from rhizome
Stems: light green to light blue-green;
Leaves: grayish-green to greenish-blue; up to 30 cm long, 6-12 mm wide; drooping, rough margined, hairless, folded, taper to a point; ligule reddish brown
Infloresence: terminal; feathery, up to 15 cm long; pale brown to grayish green
Spikelet: single flowered: first glume one-veined, second glume three-veined
Floret: lemma less than 0.5 mm
Habitat: moist woods, edges of swamps, ponds, floodplains
Similar Species: Drooping Woodreed, *C. latifolia*, has more drooping leaves and infloresence, the spikelet has a second glume with only one vein, and the ligule is colorless

Eastern Bottlebrush Grass

Elymus hystrix L. *hystrix*

Status: Unknown
Poaceae

Plant: small tufts or solitary stems
Stem: 60-150 cm
Leaves: sheaths smooth to rough or hairy; flat, 8-15 mm wide, short ligules
Infloresence: 5-12 cm, open; usually 2 spikelets at each node; rachis two-edged, internodes 4-10 mm
Spikelet: horizontal when mature; usually 2-flowered, disarticulating above the glumes and between the florets; glumes lacking or narrow and awn-like, up to 16 mm long
Floret: lemma smooth to hairy, straight, body 8-11 mm gradually tapering to rough straight awn 10-40 mm long; palea 8-10 mm, awnless
Habitat: moist woods
Similar Species: Virginia Wild Rye, *E. virginicus*, has well developed glumes, 10-30 mm by 0.8-2.0 mm, with expanded, bowed-out bases; Streambank Wild rye, *E. riparius*, also has well developed glumes, 15-30 mm by 0.4-1.0 mm, but they are not expanded nor bowed-out at the base

Riverbank Wildrye

Eragrostis frankii C.A. Mey. ex Steud.

Status: FACW
Poaceae

Plants: annual; growing in low dense tufts
Stems: 10-50 cm, repeatedly branched
Leaves: sheaths hairless, middle sheaths longer than internodes; blade 4-10 cm long by 1-4 mm wide; ligule a band of short hairs
Infloresence: 5-20 cm; ellipsoidal with middle branches longer than upper or lower; lower branches usuallly solitary
Spikelet: 3-6 flowered, 1.0-2.5 mm wide; 1st glume 1.0-1.5 mm, 0.75-1.25 x length lowest lemma; 2nd glume 1.0-1.8 mm
Floret: lemma 1.1-1.6 mm, deciduous from rachilla, 3-veined, awnless; palea 1.0-1.5 mm, persistent
Habitat: riverbanks, sandbars, moist ground, forest openings

Rattlesnake Mannagrass

Glyceria canadensis (Michx) Trin.

Status: OBL
Poaceae

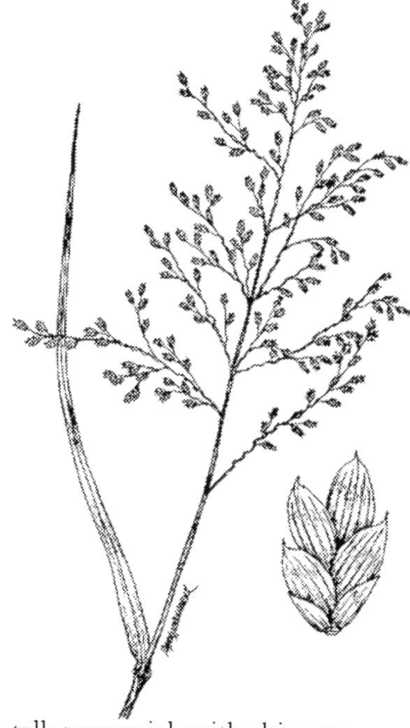

Plants: up to 100 cm tall; perennial; with rhizome
Stems: solitary or a few in a tuft
Leaves: 2-5 mm wide; ligule 2-6 mm
Infloresence: drooping, 10-30 cm; branches with spikelets mostly on ends
Spikelets: 4-10 flowered; broadly ovate, 4-8 mm by 3-4 mm; second glume longer than first; greenish
Florets: broadly ovate, lemma 2.9-4.0 mm long, flattened from side to side, awnless, usually seven veins
Habitat: shallow water at edges of lakes and ponds, swamps, fens
Similar Species: Northern Mannagrass, *G. borealis*, has spikelets than are cylindrical rather than flattened; in all the others the spikelet is flattened but usually only 2.0-2.5 mm wide; Northeastern Mannagrass, *G. melicaria* ,has a very stiff erect infloresence; in the next two the spikelets are usually purple-green: in Fowl Mannagrass, *G. striata*, the spikelets are 2.5-4.0 mm long, while in American Mannagrass, *G. grandis*, the spikelets are 4.0-6.5 mm long with membranous glumes

Sweetgrass

Status: FACW
Poaceae

Heirochloe odorata (L.) P. Beauv.

Plants: 30-60 cm tall; in small erect clumps; perennial; creeping; usually a single stem along with several blades growing from a rhizome; often clumps lack an infloresence and reproduce asexually
Leaves: 10-30 cm long; broad, hairless; dark shiny-green upper surface; produce a sweet vanilla-like odor when crumpled
Infloresence: pyramidal; 5-10 cm
Spikelets: the lower two florets are staminate; the upper one bisexual; the glumes are broad and thin-textured, 4-6 mm, the second one slightly wider
Florets: lemmas awnless; lemmas from staminate florets 4-6 mm (as long as glumes); lemmas from bisexual florets shorter
Habitat: moist soils of wet meadows & prairies, around marshes and bogs, streambanks

Rice Cutgrass

Leersia oryzoides (L.) Sw.

Status: OBL
Poaceae

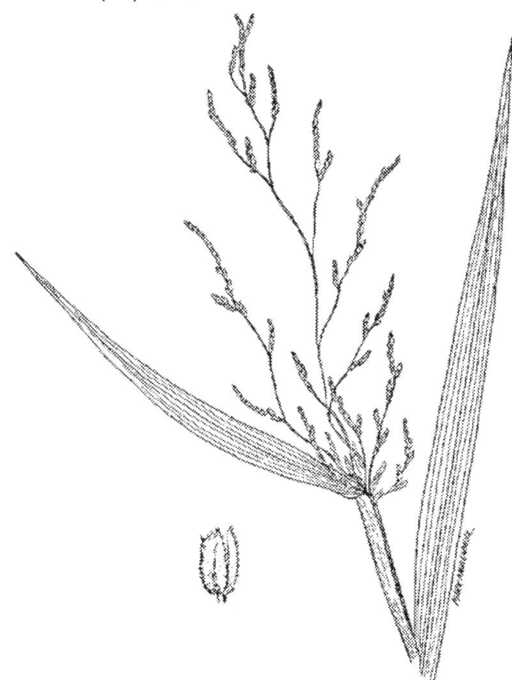

Plants: bases sprawl on ground, rooting at nodes; distally erect, to 15 cm
Leaves: sheaths with coarse backward facing teeth; blades rough margined, 15-30 cm by 6-15 mm
Infloresence: greenish to whitish panicle, exserted or partly enclosed (late season); lower part of panicle with 2 or more branches per node
Spikelet: ascending, 4.0-7.5 mm, arranged in overlapping spike-like clusters of 3-8 at end of branches; single fertile flower; laterally compressed; disarticulating at a small cup-shaped structure on the pedicel; glumes absent
Floret: lemma awned or awnless; oval, keeled, 5-veined with more conspicuous veins near the margins, few straight hairs; palea similar to lemma but narrower, 3-veined; stamens 3, stigmas 2
Habitat: swamps, wet meadows, muddy soil
Similar Species: the only species with such rough backward facing teeth on the leaf sheaths and rough leaf margins; White Grass, *L. virginica,* has only 2 stamens in the flower, lacks the backward facing teeth on the leaf sheaths and rough leaf margins, and has only one panicle branch per node

Mexican Muhly

Muhlenbergia mexicana (L.) Trin.

Status: FACW
Poaceae

Plants: 30-90 cm tall; perennial; spreading from rhizomes
Stems: erect; branches; hairy internodes
Leaves: elongated, flat, 5-15 cm by 2-6 mm, smooth.
Infloresence: terminal on main stem and also on leafy branches; green to purplish green; panicles 7-21 cm, slender - less than 2.5 cm wide
Spikelets: closely clustered;1-flowered, purplish green; 1.7-4.4 mm; glumes narrow
Florets: lanceolate lemmas, 1.5-3.4 mm; tapering to a point at the tip; sometimes with an awn; callus with long hairs
Habitat: swamps, prairie, wet meadows, streambanks
Similar Species: Marsh Muhly, *M. glomerata*, has glumes with stiff awn tips that extend well beyond the lemma; *M. uniflora*, often sprawling and forming secondary roots, grows in tufts, and has a thicker, more open, panicle

Canary Reedgrass

Phalaris arundinacea L.

Status: FACW+
Poaceae

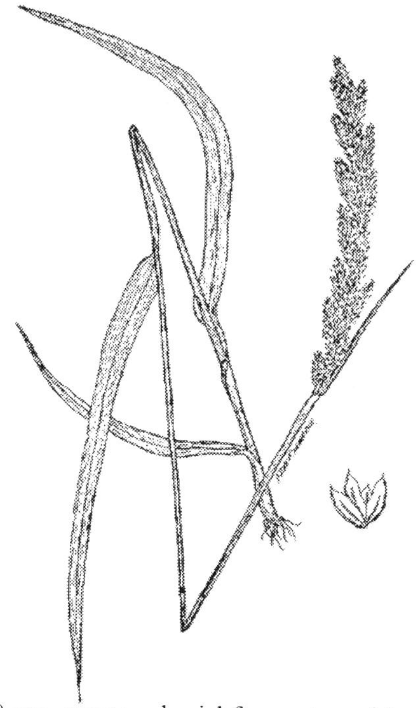

Plant: stems 70-150 cm, stout; colonial from stout rhizomes
Leaves: wide leaves, main blades usually 10-20 cm by 10-15 mm; ligule membranous
Infloresence: single terminal panicle 7-25 cm; branches usually further branched; often with light purple tinge when young; branches spread open at flowering, then closes up as it matures changing the overall appearance of the infloresence
Spikelet: greenish to purplish when young, then tan at maturity; one bisexual floret above two staminate florets; disarticulating above the glumes; glumes unequal length, compressed, keeled, 4.0-6.5 mm, wingless, smooth to rough
Floret: lemmas without awns; 2 sterile lemma at base, 1.0-2.0 mm; 1 fertile distal floret, lemma 3.0-4.5 mm, appressed hairs on distal end
Habitat: streambanks, lakesides, marshes, moist ground

Common Reed

Phragmites australis (Cav.) Trin. ex Steud

Status: FACW
Poaceae

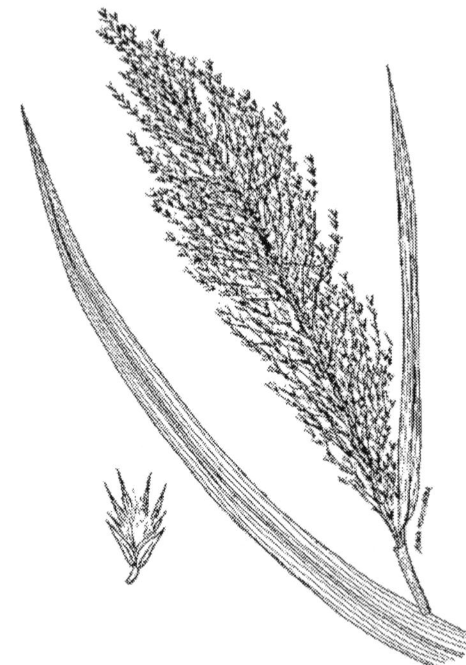

Plants: tall, culms 2-4 m; stout; extensively colonial from stout rhizomes
Leaves: ligule 1 mm; sheath open; leaves flat, long sharp pointed; 2-3 cm wide, 15-40 cm long
Infloresence: terminal much-branched plume-like panicle, 15-35 cm long; often purplish when young and straw-colored at maturity
Spikelet: 3-10 florets, disarticulating above glumes and at base of rachilla joints between lemmas; rachilla densely hairy (as long as lemma) above first floret, exposed as lemmas diverge during flowering; glumes 3-veined, lower 3-7 mm, upper 5-10 mm
Floret: lemmas narrow, 8-15 mm (upper in spikelet usually shorter), 3-veined; palea 3-4 mm, membranous; anthers 1.5-2.0 mm, purplish
Habitat: swamps, wet shores; tolerant of salt
Note: most of the Common Reed found in the Adirondacks are the exotic "Haplotype M", which have (amongst other differences) dark gray-green leaves and a dull, rough, tan stem during the summer rather than the yellow-green leaves and shiny, smooth reddish-chestnut stem of the native form

Fowl Bluegrass

Poa palustris L.

Status: FACW
Poaceae

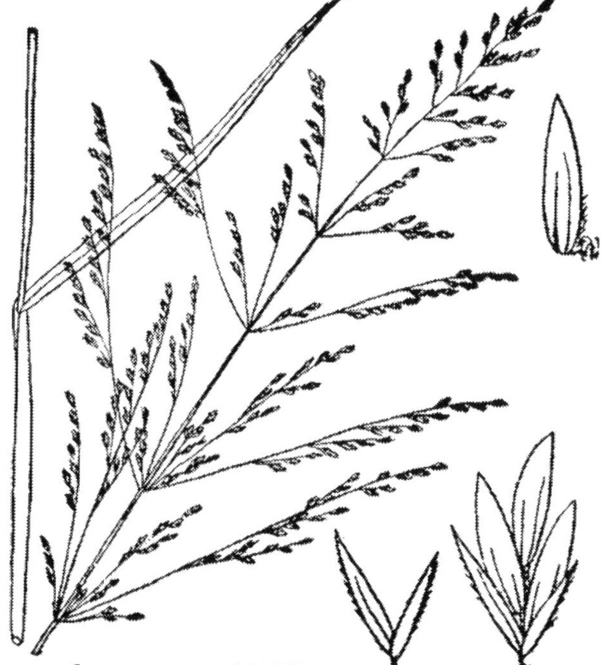

Plants: clumps of stout stems; 50-150 cm
Leaves: linear, 5-20 cm long, 1-2 mm wide; flat; ending with a boat-shaped tip; ligule ovate or triangular, 2.5-5.0 mm
Infloresence: solitary, often pointed downward or drooping; 2-5 branches at each node of the panicle; numerous spikelets beyond mid-point of branch
Spikelets: 2-4 florets, upper florets may be staminate or without flowers; glumes lance-shaped, the first glume 1.9-2.7 mm, the second 2.0-3.1 mm
Florets: lemma keeled with 3 (5) additional veins, no awns; tuft of cob-web-like hairs on lower ends of all veins, including marginal ones
Habitat: wet meadows, swamps, streambanks
Similar Species: *P. alsodes* is similar but the marginal veins of the lemmas are hairless; Fowl Mannagrass, Glyceria striata, is superficially very similar but the lemmas do not have cob-web-like hairs at the bases of the lemmas

Freshwater Cordgrass

Spartina pectinata Bosc ex Link

Status: OBL
Poaceae

Plants: stems 1-2 m tall, green to greenish brown; unbranched; with coarse rhizomes
Leaves: flat; green to yellowish green; up to 90 cm long and 2.5 cm wide; curve upward then downward and taper to a long slender point; leaf sheaths finely ribbed and glabrous; ligule at the junction of the blade and sheath with a ring of dense white hairs
Infloresence: 20-40 cm long; with many ascending branches; spikelets crowded, arranged along one side of branch like teeth on a comb
Spikelets: green and glabrous and later turn brown after blooming; strongly flattened; a pair of glumes and a single floret; glumes linear to lanceolate; first glume 5-13 mmm long, 2nd glume 10-13 mm long with rough midvein prolonged into a 3-10 mm awn
Florets: linear-lanceolate, awned; lemma and palea ~ 10 mm long, linear to lanceolate
Habitat: wet meadows, swamps, prairie, riverbanks, edges of lakes and ponds

Annual Wildrice

Zizania aquatica L.

Status: OBL
Poaceae

Plants: 2-3 m tall; annual; erect, emergent
Leaves: 90-120 cm long, 2.5-5.0 cm wide; flat, smooth
Infloresence: erect, terminal; up to 60 cm long by 30 cm across; branched - upper branches staminate, often with a few abortive pistillate spikelets; lower branches pistillate
Spikelets: unisexual, no glumes; staminate spikelets on erect upper branches, drooping at flowering, 6-11 mm, purplish to straw-colored; pistillate spikelets on drooping lower branches, bright yellow-green when ripe
Florets: lemma of staminate florets awnless or with awn to 3 mm; lemma of pistillate florets 3-ribbed, dull rough, tapering into an awn up to 7 cm long
Habitat: shallow waters (up to 1 m) in marshes and ponds, wet meadows, streambanks
Similar Species: the staminate spikelets grouped above the pistillate spikelets distinguish this species from others

Index

Illustrations for the following graminoid species are, in part, uncopyrighted material from the U. S. Department of Agriculture website (www.usda.gov):

Juncus bufonius
Juncus effusus
Juncus gerardii
Juncus acuminatus
Juncus pelocarpus
Scirpus atrovirens
Scirpus atrocinctus
Scirpus cyperinus
Schoenoplectus subterminalis
Schoenoplectus tabernaemontani
Schoenoplectus americanus
Bolboschoenus fluviatilis
Eriophorum virginicum
Eleocharis acicularis
Eleocharis compressa
Eleocharis palustris
Eleocharis robbinsii
Rhynchospora alba
Cladium mariscoides
Cyperus strigosus
Dulichium arundinaceum
Carex disperma
Carex vulpinoidea
Carex stipata
Carex trisperma
Carex atlantica
Carex projecta
Carex prasina
Carex flava
Carex lasiocarpa
Carex crinita
Carex torta
Carex aquatilis
Carex stricta
Carex hystericina
Carex lurida
Carex retrorsa
Carex vesicaria
Carex lupulina
Agrostis gigantea
Bromus ciliata
Calmagrostis canadensis
Cinna arundinacea
Elymus hystrix
Eragrostis frankii
Glyceria canadensis
Heirochloe odorata
Leersia oryzoides
Muhlenbergia mexicana
Phalaris arundinacea
Phragmites australis
Poa palustris
Spartina pectinata
Zizania aquatica